"I cannot imagine a better companion for understanding South Africa's magical bushveld." – Don Pinnock, investigative and nature journalist

"Written essentially for the lay-person Bruce has encapsulated the essence of what makes the bushveld tick!" – Dr Eugene Moll, Extraordinary Professor, Department of Biodiversity and Conservation, University of the Western Cape

"A pulsating journey through bushveld ecology. Dr McKenzie's ecological companion is a wonderful springboard into a deep understanding of how energy is used by and flows through bushveld organisms." – Dr Anthony Mills, Extraordinary Professor (Soil Science), University of Stellenbosch

"I would like to thank Dr Bruce Mckenzie for developing one of the most accessible books that encapsulates the essence of ecosystem functioning. Moreover, the intention to make this well-written book affordable to all, is commendable." – Dr Razeena Omar, Chief Executive Officer, CapeNature

An
Ecological Guide to the Bush

Bruce Mckenzie

First published by Jacana Media in 2022

10 Orange Street
Sunnyside
Auckland Park 2092
South Africa
+2711 628 3200
www.jacana.co.za

© Bruce Mckenzie, 2022
© Cover photograph: Gallo images

All rights reserved.

ISBN 978-1-4314-3287-5

Cover design by Maggie Davey and Aimèe Armstrong
Editing by Megan Mance
Proofreading by Lara Jacob
Set in Sabon 10/14.5pt
Printed and bound by Tandym Print
Job no. 003959

See a complete list of Jacana titles at www.jacana.co.za

Supporters

DONORS
Mapula Trust

Nicholson Trust

Robert Smith

SPONSORS
Erica Bester

Brian Christie

Wiley Family

Nathan & Hilary Finkelstein

Susan Gie

Tim Hoffman

Keith Kirsten

Paul & Linda Loffler

Simon Lorentz

Alice & Eugene Moll

Dion O'Cuinneagain

Caroline Petersen

Julia Price

Clive Thompson

Louise Wiseman

Ted Woods

SUBSCRIBERS

William Bond

Pat & Mike Buchel

Jonathan Dawson

Errol Douwes

John Green

Bernd & Nicola Julicher

Rudi Loutit

Andrew Mckenzie

Kenneth Mckenzie

Malcolm Mckenzie

Andy Moore

Marinda Nel

Nick Pickard

Brendan Smith

Pete Sutherland

Frank Webb

Zimbali Estate

OTHER SUPPORTERS

Felicity Mckenzie & daughters

Ndumu Club

Contents

Dedication	9
Acknowledgements	11
Foreword	13
Preface	17

SECTION A: INTRODUCTION TO ENERGY IN AN ECOSYSTEM

Chapter 1: Energy flow and transfer	23
Chapter 2: Energy requirements related to size and shape	29

SECTION B: PRIMARY PRODUCERS: PLANTS

Chapter 3: Description of savanna (bushveld)	35
Chapter 4: Changes in two palatable tree species along the east-west rainfall gradient	45
Chapter 5: Ecological challenges related to leaf size and shape in hot, dry areas	49
Chapter 6: Palatability and acceptability of graze and browse	55

SECTION C: ANIMALS AND ENERGY: THE BASICS

Chapter 7: Thermoregulatory mechanisms: Endothermy and ectothermy	65
Chapter 8: Shape and size considerations	69

SECTION D: PRIMARY CONSUMERS: HERBIVORES

Chapter 9: Who are the herbivores?	75
Chapter 10: Endothermic herbivores: Mammals	79
Chapter 11: Endothermic herbivores: Birds	99
Chapter 12: Ectothermic herbivores: Reptiles	109
Chapter 13: Ectothermic herbivores: Invertebrates	113

SECTION E: ANIMALS THAT FEED ON ANIMALS: SECONDARY CONSUMERS AND BEYOND

Chapter 14: The predators of animals	121
Chapter 15: Endothermic predators: Mammals	125
Chapter 16: Endothermic predators: Birds of prey	133
Chapter 17: Endothermic predators: Insectivorous birds	137
Chapter 18: Ectothermic predators: Reptiles	141
Chapter 19: Ectothermic predators: Invertebrates	153

SECTION F: THE DECOMPOSERS: THE FINAL CONSUMERS

Chapter 20: Decomposition: The process	159
Chapter 21: Decomposers of dead plant parts	161
Chapter 22: Decomposers of the bodies and waste products of consumers	165

SECTION G: REFLECTIONS

Chapter 23: Consolidating the theme	171
Appendix 1: Determination of ecological capacity of defined savanna ecosystems	177
About the author	187
Further readings	189
Index	191

Dedication

This book is dedicated to the memory of three individuals who, over many years, contributed to advancing my interest in the bush. My best friend and cousin, Simon "Jack" Mckenzie, a layman in conservation terms, shared many bushveld adventures with me from a young age and was very encouraging and supportive in the development of, and approach used in, this companion. Then there were two legends in the field of conservation who inspired me over the years with their knowledge and understanding as to what makes the bushveld tick and their novel ideas to the conservation thereof. Garth Owen Smith, whose career focused on the arid bushveld, and Jim Feely, who knew the moist bushveld intimately, were conservation visionaries who not only inspired me through hearing and reading about their experiences and approaches but also from the privilege of having been their companion in bushveld adventures. Both were usually quiet, with reserved demeanours, seldom arrogant and never adopting a superior attitude, to which they were certainly entitled. The times spent in the bush with them were crucial to informing my own understanding and approach to ecology, the teaching thereof and, indeed, life in general. Although both were extremely competent in their knowledge of bushveld species, their focus in discussions usually centred around the importance of understanding bushveld ecology. This understanding included the place of humans and the responsibilities that rest with them if the bushveld is to be conserved for future generations.

Acknowledgements

A sincere thank you to those people who helped in various ways with the development and production of this book. Firstly, I acknowledge the interactions with students and peers who, over many years, in no small way, contributed to the approach used in the book. I am particularly indebted to Professor Eugene Moll who, after agreeing to view and critique an initial draft, enthusiastically championed the subsequent development of the book through to its final production. My friends and family who have shared many bushveld experiences with me were resolute in their support. A special thank you to my wife, Felicity, who spent many hours suggesting changes and editing earlier drafts. Meg Jordi, the illustrator, a big thank you for the time spent in adjusting and perfecting some of my rather vague requests.

Thank you to Dr Dave Gwynne-Evans who enthusiastically agreed to partner his organisation CASABIO in the project. A big thank you to Drs Razeena Omar, Anthony Mills and Eugene Moll, and journalist Don Pinnock, who all agreed to champion the publication after perusing an early draft. Finally, thank you to Carol Broomhall and her team at Jacana for enthusiastically supporting the concept from the beginning and professionally handling all the publication processes.

Foreword

I first met the young Bruce McKenzie when I came to the University of Cape Town's Botany Department in mid-1973, when he was in his second year. Back then he was 'one of the boys'. He was bright and perhaps a little lazy, a keen and good sportsman, and enjoyed drinking beer in the Pig-n-Whistle with his mates!

As he quickly matured, he became much more focused on his studies and began to develop his full potential – but critically his love-of-life and his fellowship was not diminished. As such he became the ideal role model – and his leadership style in the South African conservation community was key to a most successful career.

Bruce has always been an independent thinker with an excellent world view. When he presented a paper at a national Grassland Conference on some of his PhD research, he realised that an academic research career was not for him and he subsequently focused on the teaching aspect, instilling confidence in, and creating opportunities for, interested students.

Bruce has always been a dedicated family man (he is married to Felicity, they have three daughters and one granddaughter – all living in Cape Town) with a love of community, which he served through Round Table, but most importantly as an educator. In this role he moulded a life based on his passion for conservation and, most importantly, working with previously disadvantaged students, firstly at UNITRA [now Walter Sisulu], then at the University of the Western Cape, and as the first CEO of the Botanical Society of South Africa. When he retired in 2008, he was presented with an opportunity to continue teaching and further support young people in the conservation arena, so he accepted an appointment in the Conservation Department at the Cape Peninsula University of Technology where he was primarily responsible for the Extended Curriculum Programme until 2016. He was awarded the university's Excellence in Teaching accolade and prize for his contributions. Bruce was involved with influencing and providing confidence to no fewer than twenty postgraduate students during his academic career. These included six who went on to complete their doctorates and all of them hold executive positions in academia or parastatals. Most of the others obtained Master's degrees and are employed in senior and middle management positions in the conservation or environmental education sectors.

Bruce's passion has always been to provide key ecological information as simply as possible, so that we, the people, can better understand the world in which we live and so make a difference in the future. I am firmly of the opinion that this book will go a long way in furthering that passion.

Written essentially for the lay person, he has encapsulated the essence of what makes the bushveld tick. Not only has he achieved this, but he is determined to have the book readily available to anyone who works in nature at minimal cost. As such he will distribute some books free-of-charge to the most deserving South Africans, and at a greatly reduced price to many more.

For me, knowing Bruce and his family over many decades has enriched my own life and passion for learning and sharing

Foreword

knowledge with our citizen scientists. His book gives the basic principles of how ecosystems function and will enable the coming generations to better understand, enjoy and thus conserve our uniquely South Africa natural heritage. He is one of my unsung heroes.

DR EUGENE MOLL
EXTRAORDINARY PROFESSOR, DEPARTMENT OF BIODIVERSITY AND CONSERVATION, UNIVERSITY OF THE WESTERN CAPE

Preface

The idea of this guide came about 15 years ago, after family and friends suggested that I provide them with some ecological background for their trips to the bush. Initially, I was rather intimidated by the request as I considered myself, at best, an arm-chair bushveld/savanna ecologist. Most of my lecturing experience and research had been in plant ecology and concentrated in the southern regions of South Africa where the main vegetation types were fynbos, forest, grassland and karoo. However, towards the end of my lecturing career, I had an opportunity to gain an understanding of the savanna (often referred to as the bush or bushveld) in a bit more detail, as I lectured nature conservation courses, which included animal ecology. I also travelled to the bushveld more often, which enabled me to become more familiar with the savanna ecosystem. I spent 10 years at the Botanical Society of South Africa, which stimulated my interest in amateur naturalists, and this guide has thus been compiled with the citizen scientist and prospective nature conservationist in mind.

One of the intriguing challenges of being a lecturer is that one is often surprised with how much you do know but of equal importance is the realisation that there is a lot that you do not know. Even when you do know things, the ability to communicate this to learners and non-biologists is often a challenge on its own. Certainly, in my latter years of lecturing, I attempted to instil in students the need to question and think about issues from different angles. In the biological discipline there are invariably exceptions to the rule, which should create opportunities for lateral thinking.

I often referred to the idiom 'as dead as a dodo' in my introductory lecture. Most students had knowledge of this phrase and its meaning from their English-language classes at school, but they were oblivious of its ecological interpretation and were fascinated when I told them the ecological narrative.

The dodo was an awkward-looking bird only found on the island of Mauritius, which was flightless and turkey-sized with no predators. Its demise came at the hands of sailors whose ships stopped at the island on their journey between Europe and the Far East. The dodo was an easy catch and a delicious meal, so much so that it became extinct some three hundred years ago.

I found it useful to consider the Afrikaans-language idiom 'so dood soos 'n mossie' (as dead as a sparrow) in a similar way. This phrase was familiar to me growing up on a Free State farm, where the common sparrow regularly perched on farm fences. It was an easy shot for a young lad, either by means of a catapult or pellet gun. Many dead but not extinct sparrows were the result of this action.

The idioms from the two languages have similar meanings but the roots are different. They differ in that the one refers to a rare bird that is now extinct and the other uses a bird which is widespread in South Africa as its point of reference. I used these idioms as an introduction to the ecological concepts of rarity and commonness, trying to get students to look at concepts from different angles and think out of the box. Some students would ask the question: Has the island ecosystem been affected by the loss of the dodo? I then had the chance to explain the importance of interconnectedness in ecosystems. An important timber tree on the island was showing little regeneration, even though it was still producing seeds. One researcher suggested that the reason for the lack of regeneration was because its fruits were an important energy source for the dodo and seeds contained in the fruit would not germinate unless their coats had been scarified by the action of the dodo's gizzard. This explanation, and the research supposedly supporting it, has been challenged in some respects but nevertheless is an important illustration of the interconnectedness of species in an ecosystem and how easily the human species can interfere with ecosystem functioning.

I adapted my focus over the years from one of concentrating on identification and classification, to one of considering energy input and output, not only to and from the ecosystem, but also between that of individual plants and animals. Two other idioms from the English language are of relevance here. I distinctly remember my mother saying to my brother and me, 'Why does one of you eat like a horse, while the other eats like a bird?' Being of farming stock, we both knew the horses always seemed to be grazing continuously while most of the birds would appear to nibble at some small item, such as a seed. We could never agree as to which of us ate like a horse!

It was only later, in my pursuit of understanding ecosystems, that I started to delve into the significance of the sayings. Of course, a horse and its close relative, the zebra, must eat a lot more than a small bird, a simple rule based on the size of the animal. Not factored into this, however, were questions such as what type of food they eat and how they have adapted to ensure that the relative amount of energy supplied by their food preferences supports their activities. For example, is it relatively more energy-sapping for a bird to fly than for a zebra to gallop?

This type of question intrigued me, and I started to consider the size and shape of organisms and their activity patterns, which would lead back to the quality and quantity of energy (food) that would be required to support their lifestyles. It is these issues on which I concentrate in this guide, as an introduction to a basic understanding of the ecology of the bushveld. Discussion on water bodies and their associated flora and fauna is not covered, nor is any in-depth description of fire ecology.

Ideally, this guide can be consulted at home or, when on safari, retrieved from the vehicle cubby hole during those quiet times, waiting at a water hole or back at camp. This guide, while hoping to increase the reader's ecological understanding, also creates the space for problem solving, critical thinking and creative engagement. Enjoy.

SECTION A
Introduction to Energy in an Ecosystem

This section describes some of the basic concepts that the reader needs to be familiar to better understand energy within an ecosystem.

The first chapter considers the flow of energy within a defined terrestrial system. It traces the inflow of energy from the sun, through the various trophic (feeding) levels and the return of all the incoming energy across the ecosystem boundary. The description includes references to basic chemistry and reactions involved in energy flow to set the scene for later sections. It also introduces the importance of the two main thermoregulatory mechanisms used by organisms, because they are determinants of energy requirements and are important in discussions of organism size and shape.

The second chapter provides information on why organisms of different shapes and sizes can influence energy expenditure and requirements. It basically considers energy transfer within and between organisms and the whole system. It also provides the background to more detailed size and shape examples discussed in subsequent sections, which follow energy flow through the different feeding levels.

CHAPTER 1
Energy flow and transfer

Energy flow through an ecosystem adheres to the two laws of thermodynamics. The first law states that energy cannot be created or destroyed, and this explains why ecosystems need a continual source of energy, usually from the sun. The second law states that some energy at each transformation is degraded into a less available form, such as heat. For example, plants capture energy from the sun during photosynthesis, but during respiration, they return part of this energy across the ecosystem boundary as heat. The remaining energy present in plants after respiratory loss represents the net energy captured. It is retained in the plant bodies, mainly in the form of carbohydrates, proteins and fats.

This energy, the net primary production, flows through the ecosystem by passing through successive trophic levels of consumers (herbivores, carnivores, omnivores and decomposers). These consumers use the energy in foods for their own metabolism and, in the end, all energy is released through respiration and returned across the ecosystem boundary. Thus, no energy is gained in the defined ecosystem; whatever energy comes into the ecosystem eventually leaves again.

Readers may be familiar with energy pyramids, which are often used to indicate the amount of energy in each feeding (trophic) level, with a figure of 10% being used as the energy available for consumption by each successive feeding (trophic) level. This can be a bit misleading, because the all-important decomposers (consumer organisms that break down dead organic material) are often excluded from the pyramid. In a terrestrial system such as the savanna, the first trophic level would include energy present in the leaves of plants, which has been captured from the sun. This primary net plant production energy would be represented at the base of the pyramid. The second trophic level would represent the proportion of energy captured by plants that is actually consumed by the herbivores (the primary consumers) and subsequent levels would indicate the amount of energy captured by secondary and higher order consumers. This avenue for energy flow is usually referred to as the grazing food chain and excludes the energy flowing through decomposers, which feed on dead plant and animal tissues from each trophic level.

The latter process is referred to as the detrital food chain. However, this food chain should not be separated from the grazing food chain, as there is too much overlap. For example, birds in the grazing food chain feed on termites and earthworms, which are essentially decomposers and thus part of the detrital food chain. Certainly, in terms of energy use and throughflow, all the organisms within a defined ecosystem must be considered as part of one system, as all contribute to the eventual loss of all energy across the defined boundary, through heat released during respiration.

Respiration is a chemical reaction whereby energy is released, usually from glucose, for use by all body cells. This is demonstrated by the reverse equation of photosynthesis, where glucose, the carbohydrate product of photosynthesis, in the presence of oxygen, produces carbon dioxide, water and energy.

$$C_6H_{12}O_6 + 6O_2 = 6CO_2 + 6H_2O + \text{ENERGY}$$

The energy released in respiration, in both plants and animals, has two components: heat (about 60%) and Adenosine Tri-Phosphate (about 40%). Heat is used to warm the bodies of certain groups of consumers, in order to maintain a steady body temperature, and any excess is released from the body and returned (lost) across the ecosystem boundary. Adenosine Tri-Phosphate (ATP) supplies the energy needed for an organism to grow and to perform activities such as muscle action and digestion. All the chemical processes that occur in the body are referred to as metabolism and the sum of these processes over a period of time is referred to as metabolic rate. Metabolism and the metabolic rate of organisms will be discussed regularly throughout this book.

ATP is primaily sourced from carbohydrates, although proteins and fats provide additional sources, if required. These three main food sources are, of course, broken down during digestion, so they can be transported across the gut wall into the body. The breakdown of soluble carbohydrates, such as starch and sucrose, is achieved quite easily by means of enzymes, whereas the break down of structural carbohydrates, such as cellulose (fibre in plant leaves), is quite a complex process. To put all this into perspective, a cellulose chain may contain between 2 000 and 14 000 glucose molecules, while starch may contain between 500 and 2 000 000 glucose molecules; both of these are therefore crucial sources of energy for organisms.

Proteins, in the main, are broken down to amino acids and these are used to make new proteins, which in turn can be used for purposes such as building new muscles. However, amino acids may be used to produce glucose if required. Fats are mainly used as an energy store and only used when the rate of carbohydrate breakdown is slow and cannot keep up with energy requirements. Fat contains more than double the energy content of proteins and carbohydrates and is thus liberally stored by animals that move regularly and fast. Migratory birds, for example, can store fats of up to 60% of their body weight. Fat molecules are continously transported from reserve sites to muscles and used as fuel in flight, for hours or sometimes even days.

BOX 1: THE IMPORTANCE OF ATP AS AN ENERGY SOURCE

The importance of ATP as a source of energy is indicated by the fact that metabolism in a typical cell uses about 10 million ATP molecules per second and recycles all the ATP within 20 to 30 seconds. To deliver all the energy stored and used by ATP, there needs to be a continual supply of glucose to the cell. During aerobic respiration, when oxygen is present, only about 38 molecules of ATP are produced from each molecule of glucose. Anaerobic respiration, which takes place in the absence of oxygen, is even less efficient and only two molecules of ATP are produced from each molecule of glucose. Thus over 90% more energy is available for work under aerobic conditions. Anaerobic respiration kicks in when oxygen cannot be delivered fast enough to cells. Glycogen stored at the muscle sites is broken down rapidly to glucose, which in turn is broken down to supply enough ATP molecules for work. Unfortunately, this anaerobic process is of short duration and causes a build-up of lactate, which is poisonous and slows down muscle contraction. It can, therefore, only be sustained for a short period of time, seconds rather than minutes, by all animals.

Most of the sun's energy that is captured by plants in the savanna is found in the leaves of grasses and trees, and is available for uptake by herbivores. The herbivores must break the chemical bonds of the leaf carbohydrates, proteins and fats in their digestive tracts for absorption into the body. This is a thermodynamically costly process, just as it is for the animals feeding on animal tissues in subsequent feeding levels. The activities of all animals must be supported by the energy obtained from food. The actual energy requirement depends on the thermoregulation method adopted, and the size and shape of the particular animal.

Which animals are endotherms and which are ectotherms?

All mammals and birds adopt the endothermic thermoregulatory strategy, while reptiles and invertebrates are ectotherms. Ectotherms rely on external sources of energy (e.g., the sun) to raise their body temperatures to a level that is optimal for activity. Endotherms use heat energy generated internally by metabolism to keep their body temperatures at the optimal level. Therefore, for similar-sized animals, an ectotherm will need far less energy from food sources than an endotherm, and the endotherm thus has a higher metabolic rate.

Production efficiencies of endotherms and ectotherms

Endotherms are generally larger animals than ectotherms, partly because the energetic cost of endothermy is very high at small body sizes. A small endotherm may only have a net production efficiency of about 10%. This measure is the amount of energy that is left over, after the majority of the assimilated energy (energy absorbed across the gut wall from digestion) has been used to maintain body temperature or lost as heat. A similar-sized ectotherm, on the other hand, may have a net production efficiency of over 60%. Thus, the different thermoregulatory mechanisms of organisms within a particular trophic level have an effect on the amount of energy passed on to the next trophic level. The digestibility of prey items may also influence assimilation efficiency and thus affect overall energy transfer. For example, mammal prey species can be digested more efficiently by predators than insect prey species. This is because insects have a higher proportion of indigestible exoskeleton in their bodies, compared to the proportion of indigestible hair on the bodies of mammals.

Thermoregulatory mechanisms and body structure of organisms are important for dictating their energy requirements (food). Of crucial importance to these requirements are the actual size and shape of a particular organism and these are discussed in the following chapter.

CHAPTER 2
Energy requirements related to shape and size

Organisms come in all shapes and sizes. When heat or water moves through the body surface, a smaller organism has a relatively larger surface area to facilitate this movement than a larger one. Not only is size important but body shape is crucial as well. The shape and size of an organism can be used to calculate the surface area to volume ratio. The biological significance of this ratio is that bigger and/or longer species exchange energy with the environment less rapidly than smaller and/or shorter species. This has a direct effect on the energy requirement (food) that the organism needs. Some examples of the surface area to volume ratio are provided below for organisms or parts of organisms of different shapes and sizes.

The shape that provides the most favourable surface area to volume ratio for an organism is a sphere. This shape would have a minimum of surface area, essentially reducing the evaporative surface, and a maximum internal volume, thus increasing the storage area for water. The shape that gives the best surface area to volume ratio would be the ideal. Two sphere-shaped animals are shown on the following page, one with a radius of 4 cm and one with a radius of 2.3 cm. The calculations for volume (V), surface area (SA) and the ratio of SA:V are shown.

An Ecological Guide to the Bush

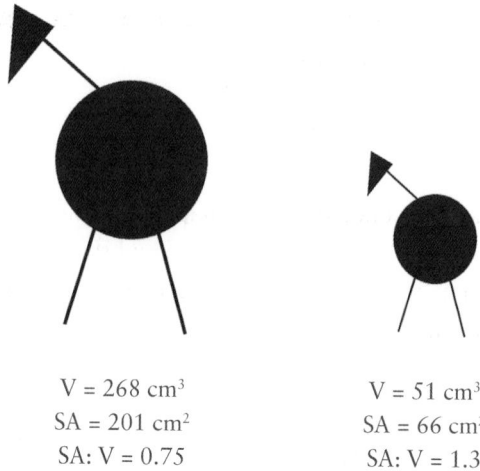

V = 268 cm³
SA = 201 cm²
SA: V = 0.75

V = 51 cm³
SA = 66 cm²
SA: V = 1.3

The following illustrations show two animals with cylinder shapes: the larger having a radius of 4 cm and a length of 4 cm, and the smaller one a radius of 2 cm and a length of 4 cm.

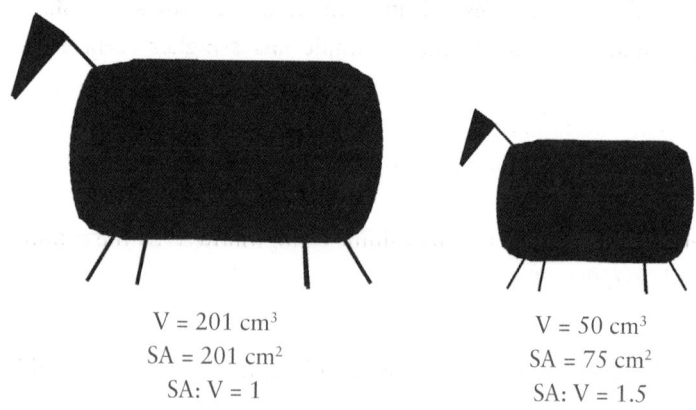

V = 201 cm³
SA = 201 cm²
SA: V = 1

V = 50 cm³
SA = 75 cm²
SA: V = 1.5

A long and thin cylinder shape (typical of snake shape) with a radius of 1 cm and a length of 50 cm is depicted below.

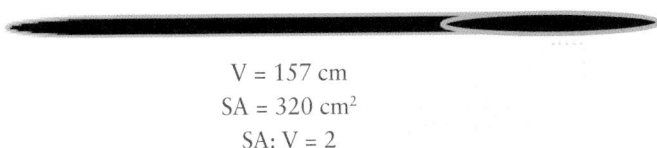

V = 157 cm
SA = 320 cm²
SA: V = 2

The final example concerns a leaf shape (ellipsoid). The larger one has a length of 20 cm, width of 5 cm and thickness of 1 cm, while the dimensions for the smaller one are exactly half of those for the larger one.

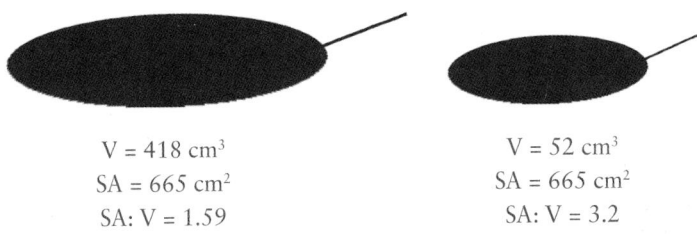

V = 418 cm³
SA = 665 cm²
SA: V = 1.59

V = 52 cm³
SA = 665 cm²
SA: V = 3.2

The above examples all indicate that, regardless of shape, the smaller and/or thinner example has a higher surface area to volume ratio, potentially exposing the smaller example to greater surface evaporation. Secondly, even where the volumes are almost equal (compare the examples which have volumes of about 50 cm³), it is shown that the sphere shape has the most favourable surface area to volume ratio, followed by the cylinder and then the ellipsoid.

Shapes, sizes, thermoregulatory mechanisms and their influence on the energy requirements of organisms will be discussed within trophic levels, beginning with the plants, the primary producers, in section B.

SECTION B
Primary Producers: Plants

Plants are the organisms responsible for capturing light energy from the sun and creating the food source (energy) for the rest of the ecosystem's organisms. The main plant organs that capture light energy through photosynthesis are the leaves – and in the savanna (bushveld) particularly, the leaves of trees and grasses.

This section concentrates on the types, composition and functional characteristics of the tree and grass components of the savanna, primarily in relation to them supplying energy for the primary consumers, the herbivores.

CHAPTER 3
Description of savanna (bushveld)

Bush and bushveld are primarily vernacular terms often used in South Africa and Namibia in place of the more scientifically accepted savanna vegetation type.

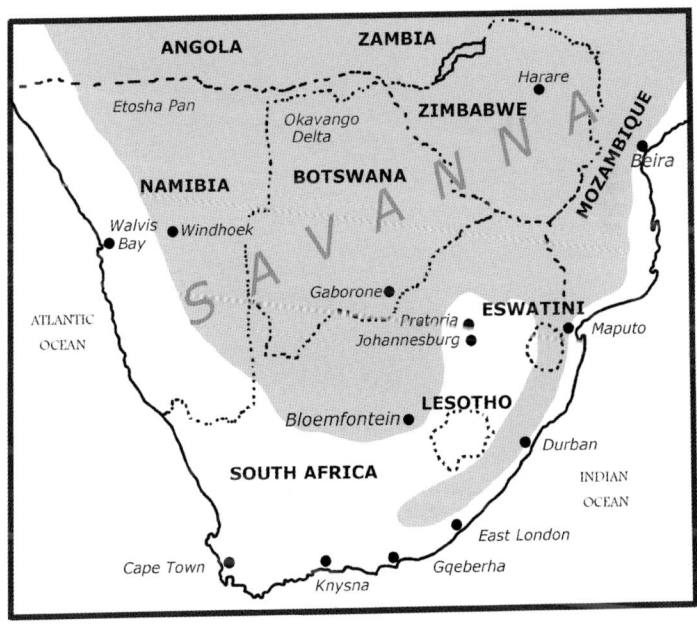

The savanna region considered in this guide is found across the north-eastern and north-central areas of South Africa, into Zimbabwe, Botswana and northern Namibia. The map on page 35 indicates the general areas in southern Africa where one can find savanna/bushveld vegetation. Savanna as described for this guide may also be found in isolated pockets outside the mapped areas, for example in the Mokala National Park near Kimberley and along or adjacent to some drainage lines in more arid areas such as in the Namaqualand in the Northern Cape Province. Savanna is thus typically found in the larger game parks and surrounds, in tropical and near-tropical areas, where the rainfall is between about 300 mm and 1000 mm per annum. Rainfall is typically concentrated in a few summer months of the year and does not supply enough water for trees to become dominant over grasses. In this guide, savanna is described as a woodland/grassland mosaic and is best represented where plant production is equally divided between an upper layer of trees and a lower layer of grasses.

There is not total agreement by ecologists as to the definition of savanna but the above description will serve the purposes of this guide. Details of the very dry savanna types on the western margin of the sub-continent are not included, but the principles as discussed are the same as for the rest of the savanna areas. A few examples from the arid savanna are, however, included, as they are often indicative of extremes and are therefore useful for an overall ecological understanding.

Tree and grass leaves supply most of the food energy for the herbivores in the savanna. The savanna areas in southern Africa have been classified into about 100 different descriptive vegetation types. Most of these descriptive types can be placed under the umbrella of two functional types or a mixture of the two – identified by scientists as broad-leafed and fine-leafed savanna types. It is these two types that are used as the basis for considering energy flow in the savanna ecosystem.

Examples of the two functional savanna types

The broad-leafed and fine-leafed tree savannas provide different utilisation opportunities for herbivores. The broad-leafed habitat tends to be favoured by trees that do not possess thorns and their leaves or leaflets are at least 1 cm in length and breadth. This type tends to be found in moister areas where the soils are leached and nutrient poor. An example from the broad-leafed type is the Red bushwillow (*Combretum apiculatum*). This commonly found tree may grow up to 20 m in height, with leaves measuring up to 3.5 x 6.5 cm in size (see diagram on page 38, which includes a flower and a few fruits).

The fine-leafed type has the opposite features, as the leaves are usually compound, and the leaflets much smaller in size than the first type. The branches are usually armed with thorns, and

Red bushwillow

they tend to occur in areas that are drier and have nutrient rich soils. A common example of this type is the Umbrella thorn (*Vachellia tortilis*). This tree grows up to 30 m tall, with leaves measuring about 3 x 1.5 mm, the smallest leaves of all the African 'acacia' species (all African acacias have recently been placed into either the *Vachellia* or *Senegalia* genera). It has thorns to deter browsing, as indicated in the diagram on page 39.

This type tends to provide the more nutritious foliage for herbivores, one of the reasons being that many of the trees belong to the leguminous *Vachellia* or *Senegalia* genera, which have a symbiotic relationship with bacteria living on their root nodules. These bacteria utilise atmospheric nitrogen (up to 200 kilograms per hectare per year (kg/ha/yr)) and incorporate it into nitrogenous compounds that in turn contribute to increasing protein levels in the leaves. The advantages of having fine (small) leaves, despite the constraints of having a high surface area to volume ratio as described in chapter 2 particularly in hot and dry areas, are discussed in chapter 5.

The grass layer under the two functional types

The nutritional differences in the tree leaf (browse) component of the two functional structural types are also evident in the leaves of the lower layer, the grass stratum (graze). The grasses in the fine-leafed type not only benefit from the richer soils, but also utilise the additional nutrients and nitrogen that is added from the decomposition of the protein-rich tree leaves. These grasses thus remain digestible throughout the year, as their protein content is always above 1%. However, grasses in the broad-leafed type have a nitrogen content below 1% in the dry season, often less than 0.5%, are fibrous and show an increase in tannin, rendering them unpalatable. In addition, where the tree component becomes denser, particularly in the moister eastern broad-leafed areas, the actual grass production becomes low and often less than 10% of it is grazed.

In the fine-leafed type, at least 30% of the grass production is grazed in any year and this could increase to 80% in good years. The fine-leafed areas thus support relatively high mammalian herbivore populations. The opposite is true for the broad-leafed savanna type and, furthermore, the trees and grasses in this type often have a relatively high concentration of unpleasant-tasting compounds, making them less palatable to mammalian herbivores.

Umbrella thorn leaf

BOX 2: SWEET-VELD AND SOUR-VELD

The more palatable fine-leafed vegetation type is sometimes referred to by rangeland managers as sweet-veld, whereas the less palatable broad-leafed type is referred to as sour-veld. Even in areas where both types receive the same rainfall, sour-veld will have a lower grazing and browsing capacity than sweet-veld, which is prone to overgrazing but may recover quickly. The division into sour-veld and sweet-veld characterise the two ends of a spectrum and there are many areas that support a mixed-veld habitat.

Herbivores that concentrate their feeding on tree leaves are referred to as browsers, while those that concentrate on grass leaves are termed grazers. Although shrubs and herbs also contribute to browse, the emphasis here will be on tree and grass leaves, because they together provide most of the energy used by herbivores.

Variations found in tree and grass layers in the hot and arid areas

As aridity increases on the western side of the sub-continent, the frequency and length of dry spells within the rainy season increases. The perennial grasses struggle in the uncertain conditions and grass productivity is extremely low. In the extreme west, in Namibia, savanna is found in the pre-Namib desert, particularly in the river valleys draining from the escarpment. The variability, in temporal and spatial distribution of rainfall, does not favour a consistent quantity of graze from year to year. Consequently, most grazers, including the large white rhino, the medium-sized Cape buffalo and the blue wildebeest, are absent. On the other hand, mega-herbivore browsers such as elephant, black rhinoceros and giraffe, through the mid-sized mixed feeders, such as gemsbok and springbuck, to the smaller browsers, such as duiker and steenbok, are present and fairly common in some areas. This is because plants such as the resident species of thorn tree, the ana tree (*Faidherbia albida*), and some succulent shrubs have deep roots that utilise groundwater even in years of low rainfall, so browse is usually in good supply.

More information relating to the two functional types of vegetation is presented below. I have concentrated on trees (browse component), as it is easier to observe the leaf differences in this component. Firstly, the distribution of the two types in the Kruger National Park is used as an example of the separation at a local scale.

Description of savanna (bushveld)

Distribution of functional savanna types in the Kruger National Park

The Kruger National Park (KNP) is a convenient area to view the distribution of the two main vegetation types at a local scale. The park is some 20 000 square kilometres in size, about 360 kilometres long and has an average width of about 65 kilometres. Except for the far north and far south, which have considerable areas of mixed veld, the central area of the park can be conveniently bisected along its length. The eastern half has extensive cover of nutrient-rich basaltic-derived soils, has a low average rainfall and mainly supports trees and grasses of the fine-leafed type. The western half has granite as its base and the soils developed from this geological type are heavily leached and thus relatively nutrient poor. This is the result of the higher rainfall received, and this area, not surprisingly, supports the broad-leafed type of tree.

If this relationship between rainfall and vegetation type is consistent across the sub-continent, it would be expected that there would be relatively more trees with smaller leaves on the drier western side, than in the east.

Leaf sizes of trees along the east-west sub-continental rainfall gradient

It is not surprising to find that there is a general decrease in leaf size of the commonest trees, as one moves from east to west in southern Africa. Moving from the wet east-coast forests, where trees with the largest leaves are found, one progresses through the moister parts of the savanna, where we find many evergreen and semi-deciduous trees with medium-sized leaves (e.g., the mopane [*Colophospermum mopane*] and Bushwillow [*Combretum* spp.]), to the drier and hotter areas of the arid west, where fine-leafed deciduous or semi-deciduous trees (e.g., camel thorn, sweet thorn, knob thorn and umbrella thorn) are common.

Deciduous trees shed their leaves annually for a whole season, whereas semi-deciduous trees lose their foliage for a very short period, when old leaves fall off and new leaves start growing. Most of the fine-leafed species of trees are deciduous or semi-deciduous, whereas a higher proportion of the broad-leafed species are evergreen, retaining their leaves throughout the year. Leaf loss occurs during the cold, dry winter season, when it is advantageous for trees to shut down photosynthetic activity and remain dormant until moisture and temperature conditions improve. Before losing their leaves, the trees will move nutrients to other sites such as their roots or trunk, and these can be mobilised quickly when conditions for photosynthesis improve.

BOX 3: THE UNIQUE ANA TREE

The ana tree is unique as it loses its leaves in summer. This is a tall fine-leafed tree common in drier areas and particularly conspicuous in the dry riverbeds of north-western Namibia. It was placed in the genus *Acacia* but has recently been included under the genus *Faidherbia*, partly because, unlike the acacias, this tree is leafless in summer. It can tap water from as deep as 40 m below the soil surface, which enables it to produce its leaves and flowers in the dry winter months and the nutritious pods (up to 100 kg per mature tree per year) at the end of the dry season. Even large browsing herbivores, such as giraffe and elephant, are thus supplied with leaves in the winter season and nutritious pods in spring. In addition, grasses grow well below the leafless canopy in summer, utilising the high nutrient levels in the soil. Of course, the grasses do not have to compete with the tree for surface water, as it is dormant during this time.

Although, in general, broad-leafed species are the more common functional type in the eastern side of the sub-continent and the proportion of evergreen trees is higher, there is no reason why fine-leafed deciduous trees cannot be accommodated, as described earlier about Kruger National Park. The same reasoning applies to broad-leafed trees that may be found in the dry west, but in localised areas where more water is available. Mopane thickets, for example, are found in extensive areas of north-western Namibia. It is also not unusual to find both functional types near each other. For example, in undulating areas, broad-leaved savanna is found on the sandy and leached crests while the fine-leaved type is found on the valley bottoms which have clayey soils that are nutrient rich.

There are general differences in structural features of individual species that are found along the climatic gradient from east to west. The following chapter considers changes across the moist to arid gradient in two palatable tree species which are important for herbivores.

CHAPTER 4

Changes in two palatable tree species along the east-west rainfall gradient

Two important fodder trees, one from the genus *Vachellia* and the other from the genus *Boscia*, are used to emphasise changes across the wet–dry gradient. The sweet thorn tree (*Vachellia karroo*), like many Vachellias, occurs all over southern Africa and is said to be a good indicator of sweet-veld. The shepherd's tree (genus *Boscia*) has eight species in southern Africa. Two are found across the east–west gradient, four are restricted to the moister eastern zones and two are restricted to the more arid savanna in Namibia. The species in the moister areas have large leaves while the two restricted to the western zone have small leaves. The two species occurring across the gradient are the closely related white-trunked shepherd's tree (*Boscia albitrunca*) and smelly shepherd's tree (*Boscia foetida*). The white-trunked shepherd's tree is more commonly seen and the rarer smelly shepherds' tree is divided into subsp. *foetida* on the western side and subsp. *longipedicillata* on the eastern side. The table below indicates the main differences between the eastern and western variants of the sweet thorn and the variation between two sub-species of the smelly shepherd's tree.

Table 4.1: Some characteristics of two fodder tree species in the eastern and western regions of southern Africa

	Sweet thorn		Smelly shepherd's tree	
Region	East	West	East	West
Trunk	Single-stemmed	Multi-branched	Single-stemmed	Multi-stemmed
Height	30 m	6 m	> 2 m	< 1 m
Canopy	Sparse	Dome-shaped	Sparse	Dome-shaped
Leaf retention	Evergreen	Deciduous	Evergreen	Evergreen
Leaf size	Large, 4–7 leaflets per leaf	Small, 2–3 leaflets per leaf	Long, about 3.5 cm	Short, about 1 cm or <
Thorns	Absent or small	Present	Absent	Absent but branches can be spine-tipped

The sweet thorn and smelly shepherd's tree share several adaptations to the east–west gradient. Trees on the moister east coast are tall, single-stemmed with a sparse or ill-defined canopy, have large leaves, and thorns, if present, are small. The arid, western trees are much shorter, tend to be multi-stemmed or multi-branched, have a well-defined, dome-shaped canopy with small leaves and thorns or spines are prominent. The sweet thorn, however, is also deciduous on the western side, while the smelly shepherd's tree remains evergreen. The latter has a very deep root system, extending more than 50 m underground, so it presumably obtains sufficient moisture in the hot, dry season to remain evergreen. Its leaves have a relatively high protein content (16%) and low tannin content (0.25%), so the leaves are highly palatable despite being evergreen. The dome shape is an advantage in a hot arid environment as it increases the ratio of interior volume to overall surface area of the plant. The air in the large interior is often considerably cooler than the air outside the canopy and this of course assists the tree in coping with the very high temperatures. The multi-stemmed nature of the tree is also an advantage, as the plant will not be killed by the loss of one or two stems. The evergreen nature of the relatively small shepherd's tree is a characteristic of many of the smaller woody plants (mainly shrubs up to 1 m) in drier environments. The plants cope with drought with strategies such as curling their leaves to lower metabolic activity, until conditions are again favourable.

There is an interesting difference in the response of the eastern and western sweet thorn variants to increased water supply. While the eastern variant trees invest more in growth, such as increased height and more leaves, the western variant tends to invest in defensive allocations, such as more thorns and increased concentration of tannins in the leaves, to discourage herbivores from excessive browsing. The illustrations below and on page 48 show some of the features of the two trees.

A sweet, slender thorn tree from the moist east coast, with a small canopy which is evergreen. The leaf is like the umbrella thorn leaf (page 39).

A sturdy, short, sweet thorn tree from the arid west. It is multi-branched with a well-developed, dome-shaped canopy that sheds its leaves in winter.

An example of a smelly shepherd's tree from the arid west. Short, multi-stemmed, with a dome-shaped canopy.

Branch of a smelly shepherd's tree with small fascicled leaves from the arid west

The sweet thorn and smelly shepherd's tree are both indicative of sweet-veld (fine-leafed type) and have particularly small leaves in the arid west. As discussed, herbivores would find these more palatable than broad-leafed trees, so one may then well ask: What are the relative advantages and challenges of having smaller leaves in drier, hotter areas and how important is leaf shape?

CHAPTER 5

Ecological challenges and advantages related to leaf size and shape in hot, dry areas

There are physical and physiological challenges to which leaves of different shapes and sizes need to adapt, a brief discussion of which is given below.

All organisms lose water through evaporation and there is great potential for evaporative loss in hot, dry environments. The temperature and water vapour pressure of the surrounding air, coupled with the organism's ability to reduce water loss, determine the amount of evaporative loss. The leaves of trees in the drier areas may have some waterproofing features such as a waxy surface, but the size and shape of leaves are crucial in helping to minimise water loss through evaporation. The geometric shape that gives the most favourable surface area to volume ratio is a sphere, followed by a cylinder, then a rectangular prism and finally an ellipsoidal shape (see chapter 2 for illustrations of shapes, sizes and worked examples).

No plants have spherical leaves, but in dry areas many succulent leaves tend to be cylindrical. Most tree leaves, however, have a shape close to that of an ellipsoid, which is not ideal in terms of surface area to volume ratio. Regardless of the actual shape, the smaller a leaf is, the greater this ratio becomes, thus increasing the potential for evaporative loss. So, size is also an important factor affecting leaf function, because more moisture would be lost from a leaf surface that is relatively large compared to its volume. One would think that a small ellipsoidal leaf, with its relatively large surface area to volume ratio, would in fact not be a good adaptation to reduce water loss.

To indicate the importance of the surface area to volume ratio, simulated data is presented for two tree branches that have similar leaf surface areas but very different volumes. One tree has branches containing three leaves each and the other has 34 small leaflets per branch, as illustrated below.

The dimensions and subsequent calculations used to determine the surface area to volume ratio for each branch are presented in Table 5.1. For ease of calculation each leaf is assumed to have a rectangular prism shape rather than an ellipsoid.

Ecological challenges and advantages related to leaf size and shape in hot, dry areas

Table 5.1: The dimensions and calculations used to determine surface area to volume ratio for a branch from two trees

	Three-leaved branch	34 leaflet branch
Leaf dimensions		
Length (L)	4 cm	1 cm
Width (W)	2 cm	0.5 cm
Height (H)	1 cm	0.5 cm
Surface area of leaf		
2 (L x W + L x H + W x H)	28 cm^2	2.5 cm^2
Volume of leaf		
L x W x H	8 cm^3	0.25 cm^3
Surface area of all branch leaves	84 cm^2	85 cm^2
Volume of all branch leaves	24 cm^3	8.5 cm^3
Surface area: Volume ratio of a branch	3.5	10

Despite the surface areas of the two tree branches being similar, the small-leafed branch has a volume about three times less than the broad leafed branch and thus a much higher surface area to volume ratio. How can the smaller leaves be an advantage if they potentially use so much more water in hot, dry environments? The answer can, at least partially, be explained by considering the boundary layer and its function.

What is the function of the boundary layer?

The boundary layer is a thin layer of air that forms just above the leaf surface. It is formed because the surface receives shortwave radiation energy and then re-radiates this back into the air as long-wave radiation, and this heats a thin layer of air (the boundary) above the leaf surface. The temperature of the air in this boundary layer is much higher than the air above it and this would influence the amount of water vapour being lost from the surface of the leaf during transpiration. Small leaves have a thinner boundary layer than large leaves and are therefore at an advantage as the boundary layer temperatures are much lower. Although many plants can tolerate temperatures up to about 60°C, large leaves with much thicker boundary layers and high temperatures can adversely affect the normal plant physiological processes, such as photosynthesis, particularly in hot, dry environments.

If the boundary layer is disturbed by wind (convection), the air in the boundary layer and air above it will mix. More heat is lost from the leaf as the air in the boundary layer becomes cooler because of mixing. The positive aspect of this is that the convection in fact cools the leaf surface, and many closely spaced smaller leaves are more easily disturbed by a breeze than a single larger leaf. The development of increased turbulence thus easily disrupts the thin boundary layer of smaller leaves, cools the leaves and prevents them reaching extremely high temperatures that may disrupt their physiological processes. Having many small leaves, rather than a few large leaves, is therefore a good adaptation.

The small-leafed, deciduous thorn trees are thus common in hot, dry environments and they benefit from growing in richer soils (sweet-veld areas), when compared to the broad-leafed trees common in the sour-veld areas which are moister but have poorer soils.

Regardless of whether sweet-veld or sour-veld is being considered, it is necessary to discuss the important differences between grass leaves (graze) and tree leaves (browse), which make each more desirable to grazers or browsers. The energy that the trees and grass plants have captured from the sun through photosynthesis is available to support the consumer organisms in the ecosystem (see chapter 1 for detail). Some reference has already been made to palatability, but it is necessary to look at the differences between grass and tree leaves in more detail to understand herbivore preferences.

CHAPTER 6

Palatability and acceptability of graze and browse

General characteristics of graze and browse

Regardless of leaf size and shape, there are important differences between graze and browse in terms of utilisation by herbivores. Grass cover tends to be more homogenously spread than trees, which are more widely dispersed in the environment. This is less so in the dry west, where grass production may vary considerably from year to year, as mentioned in chapter 3. Grass has a low growth form and new leaves are added at the base of the plant; this gives a grass plant an advantage over a tree, in that it can recover quite rapidly after being grazed and regain surface area. This growth form also allows rapid recovery after fires, which are necessary every few years for removal of old grass and regeneration of new grass. Fires are also not usually intense enough to affect the canopies of trees. Most trees have a medium to high growth form and new leaves are primarily added at the tips of branches; this is a slower process for leaf production than that of grasses. However, trees retain their green leaves for a longer period than grasses.

Leaves at the tip of tree branch   Grass plant rooted in soil

An understanding of the actual structure and contents of plant cells in browse and graze is important because herbivores have different diets and digestive systems to cope with the challenges and advantages of each.

Comparison of structure and content of browse and graze cells

Plant leaf cells have two important components that relate to palatability and accessibility to herbivores. The cell wall, which provides structural rigidity to the cell, is composed of three main compounds that affect herbivores: cellulose, lignin and silica. The second component is the cell content, where the quantity of tannin and protein is most relevant to accessibility, palatability and digestion.

Cellulose (fibre), the most abundant plant carbohydrate on earth, is a basic structural component of cell walls. Cellulose is, however, difficult to break down into simpler carbohydrates, such as glucose, and herbivores require special micro-organisms in the gut to produce the cellulase enzyme needed in the digestion process.

Lignin, the second most abundant plant compound, is also involved in maintaining rigidity in the cell wall but is very resistant to microbial action and therefore retards digestion.

Silica, composed of silicon and oxygen, can make up to 10% of the cell wall and, although it is useful to plants in that it can help reduce the damage that may be caused by high temperatures or drought, it is very abrasive and can wear down herbivore teeth.

Tannin, found in the cell content, affects digestion rate, as it binds with proteins, thus making it difficult for the proteins to be digested. Proteins, of course, need to be broken down into amino acids, which are small enough to be absorbed across the gut wall and into the bloodstream.

Although fat is a component of cell membranes, such as that of the chloroplast, it is not present in high quantities. It is nevertheless an important food source for animals that feed on seeds, as seeds have high concentrations of fat reserves which would normally be mobilised on germination.

Graze cells have thick cell walls, and these have high concentrations of cellulose and silica, while browse cells have thin cell walls with a lower content of cellulose and silica, but a higher content of lignin. Browse cells have considerably more protein content, but tannin levels are higher. Table 6.1 summarises the differences between graze and browse cells, which affect herbivore preferences.

Table 6.1: Main structural and chemical characteristics of graze and browse cells

	Browse	Graze	Herbivore adaptations
Cell walls			
Structure	Thin walls	Thick walls	**Browsers** can break cell walls easily
Cellulose content	Low	High	**Browse** more rapidly digested
Silica content	Low, about 3%	High, up to 10%	**Browse** more acceptable as less tooth abrasion
Lignin content	Higher in mature leaves	Low in mature leaves	**Browsers** concentrate on utilising young leaves
Cell Content			
Protein content	Usually > 10%	Usually < 6%	**Browse** much more acceptable
Tannin content	High	Low	**Browsers** actively select trees with less than 5% tannin, a level at which tannins have a minimal effect on digestion

The net result of the advantages and disadvantages is that browse is generally more nutritious and acceptable to herbivores than graze. Even when graze seems to have more advantageous characteristics, such as low lignin and tannin levels, browsers typically ignore it and will choose to feed on plant parts which are less of a challenge (e.g., young leaves of trees). Despite the advantages of browse, animals have adapted to being specialist grazers, mixed feeders or browse specialists.

A few strategies adopted by trees to counteract browsing

Although tree leaves may be the more acceptable and nutritious for herbivores, they also have counter strategies to reduce browsing. A few examples are presented below.

Some trees, such as the mopane, which have relatively high amounts of leaf protein (up to 16%) and are thus a favourite of browsers, also have a high content of tannin as a counter measure. In fact, browse plants may increase their tannin levels quite rapidly as a deterrent to excessive browsing and most browsing herbivores will preferably choose tree leaves which contain less than 5% of condensed tannin. At this level, the nutritional quality and subsequent digestion process is not affected. Trees will, however, react to browsing by increasing their concentration of tannins within a few minutes of being browsed. This discourages the browser from feeding and it will move off to another tree.

> ### BOX 4: TREE ADAPTATIONS TO BROWSE AND DROUGHT
>
> It has been suggested that when tree leaves are eaten, they emit volatile ethylene and this acts as a warning to other trees, and within half an hour the neighbouring trees increase the synthesis of tannin. Trees, such as the magic guarri (*Euclea divinorum*), may, in effect, increase their tannin levels as a response to drought rather than browsing and this also acts as an early warning system to other trees to increase tannin levels and thus discourage browsing in the stressful water-scarce period. These defensive mechanisms may prevent a single tree from being defoliated and gives it time (about 40 hours) to recover. Tree leaves also have their greatest concentration of tannins (about 10%) at 2 m above ground level, thus discouraging most herbivores from excessive browsing at this level.

It is not surprising that browsers are not often found in the same localities on successive days, but grazers may be in the same place. Browsers do produce special proteins in their saliva which help to bind the tannins before the rest of the digestible proteins enter the digestive system, but it is not necessarily a hugely successful counter strategy. The majority of browsing herbivores do not possess microbes that can either digest or deactivate tannins, so most of the tannins are deactivated in the liver or, along with lignin, are passed on to the environment in faeces and become subject to the decomposition process.

The advantages and disadvantages of evergreen and deciduous plants

One other advantage that browse has over graze is that trees retain their leaves longer into the unfavourable season than grasses. However, some trees are deciduous, and others are evergreen. Leaves on evergreen trees tend to be thicker and contain considerably greater concentrations of defensive compounds such as phenolics and tannins. On the other hand, deciduous leaves have a higher nitrogen content and are thus generally more palatable. Many browsers will therefore favour the deciduous trees in the good season but may have to revert to less palatable evergreen species in the cold and dry season, often losing condition in this period. A good example of this is the giraffe, an adaptable browser that utilises about 70% of savanna tree species in its diet. In the Kruger National Park, the *Combretum* and *Vachellia* trees supply about 50% of their food source in the good season. However, most trees of these two genera will drop their leaves during the winter months and the giraffe is forced to survive on the less palatable and less nutritious evergreen species for a few months, often losing condition.

After a good rainy season, however, deciduous species may retain their leaves for longer periods, up to 10 months of the year. This is important as evergreen biomass is generally low in savanna areas, often less than 10%, so that browsers only need to utilise browse from the less common and less palatable evergreen species for short periods.

Despite the advantages of browse over graze, about 50% of the annual savanna primary production is present in the lower grass layer. Many herbivores, particularly the more commonly encountered mid-sized ones, are content in utilising graze. The adaptations for this are discussed in chapter 10.

In chapter 5, the importance of size and shape of whole plants or parts of plants (leaves) were considered. Size and shape of animal organisms also have a major influence on activity and thus food energy requirements. Coupled to this is the need to consider the various thermoregulation mechanisms employed by animals, as these mechanisms have important consequences in terms of food energy required and their modes of feeding and digestion. Consideration of the above will enable an understanding of why birds and reptiles, for example, are not leaf eaters and reasons why insects are so successful as folivores. The first step is to focus on thermoregulation and then discuss size and shape.

Readers are reminded at this stage that this guide does not include water bodies or their biota such as fish, water birds and insects such as dragonflies. One mammal (the hippopotamus) and one reptile (the crocodile) that need large water bodies are, however, included in discussions.

SECTION C
Animals and Energy: The Basics

The previous section ended off with a summary of the important forage properties of leaves (graze and browse) that are available for primary consumers (herbivores). Before considering herbivory in more detail, it makes sense to consider energy aspects that are common to all animals in the savanna food chain. In this section, basic aspects of animal metabolism are introduced. The two main thermoregulatory mechanisms found in savanna are also discussed, as they influence the quantity of food energy required. The shape and size of an organism are also important in terms of energy gain and loss. In this regard, the concept of animal metabolic mass is introduced. This is a measure analogous to the surface area to volume ratio discussed in relation to plants and will be extensively used in subsequent sections of this guide.

CHAPTER 7

Thermoregulatory mechanisms: Ectothermy and endothermy

All animals gain or lose heat from solar radiation, infrared radiation, convection, conduction, evaporation and metabolism. An animal can adjust the energy exchange of these various pathways to warm up or cool down and thus maintain a desired body temperature. Most mammals, birds, reptiles and insects have body activity ranges between 30°C and 40°C. Controlling body temperature is important because most animals will at some stage encounter environmental temperatures (too hot or too cold) that may be lethal. The control of temperature is vital in ensuring that chemical reactions necessary for normal bodily performance take place within the required temperature ranges.

Some readers may be familiar with the terms warm-blooded and cold-blooded animals in relation to temperature control. These terms are misleading because it is better to refer to the source of heat which animals use to control body temperature rather than blood temperature. Control of the metabolic rate is important for temperature control. Metabolic rate is essentially a measure that relates to how quickly food (e.g., energy from glucose) is utilised to keep an animal's cells functioning. It differs amongst species and under varied environmental and activity patterns. Very importantly, much of the energy released during metabolism is in the form of heat (see chapter 1 for detail). This is not necessarily a disadvantage, as some animals can regulate and make good use of metabolic heat to keep their body temperatures at a more-or-less constant level. This strategy is termed endothermy and is found in mammals and birds. In these animals, the proportion of body energy consumption used in metabolism when resting is around 65%, while about 25% is used in physical activity and about 10% in the digestion process.

An alternative strategy for regulating body temperature, ectothermy, is found in plants, invertebrates, amphibia and reptiles. They rely mainly on external energy sources to warm or cool the body. The heating of the boundary layer and subsequent cooling of small tree leaves by convection, as explained in the previous section, was an example of ectothermy in action. Some 90% of animals are ectotherms. This high percentage is not surprising when one considers that most ectotherms are small organisms and are therefore able to occupy niches not available to most endotherms. For the purposes of this guide, the ectotherms discussed are in the main terrestrial. These are insects, a few other invertebrates and reptiles.

Mammals and birds can remain active when there is insufficient solar radiation to warm themselves, as they maintain a relatively constant body temperature. The costs of endothermy are, however, very high, as these animals require large quantities of food energy to sustain their high metabolic rates and keep a constant temperature. The ectotherms, on the other hand, do not depend on metabolically produced heat energy to keep their body temperatures at a constant level. They rely on external heat sources, such as basking in the sun to warm their bodies, so need comparatively less food energy than mammals or birds. For a lizard (ectotherm) and a similar-sized rodent (endotherm), the food requirement for the lizard may be only 10% of that required by the rodent. The smaller the body, the higher will the cost of endothermy be, therefore most of the smallest animals are ectotherms. Some of the very few exceedingly small endotherms ameliorate this challenge by having rounded bodies, which provide a favourable surface area to volume ratio (e.g., a rat).

How do endotherms and ectotherms utilise the energy obtained from food?

Regardless of body shape and size, it is important to understand what endotherms and ectotherms do with the energy they obtain from food. Up to 90% of food energy taken in by an endotherm may be used to produce heat to maintain their body temperatures. Ectotherms, of course, do not rely on metabolic heat production to warm their bodies, so well over 50% of food energy intake can be converted to biomass accumulation and production of offspring. For endotherms this could be lower than 10% (see chapter 1).

An interesting consequence of this difference is that, for a given area in the savanna, the annual increment of ectothermic herbivorous biomass (mainly insects) could be equal to that of the larger endotherms (mainly herbivorous mammals). The small ectotherms can thus be thought of as finding and using food energy not available to the larger endotherms and then efficiently converting this food energy into increased biomass, in their own bodies and in the production of offspring. One other advantage to ectotherms such as reptiles, is that they can live much longer than similar-sized mammals, probably because of less wear and tear on their bodies, which may be associated with high metabolism in mammals. These considerations are important in relation to understanding energy transfer between different trophic (feeding) levels in the food web and will be discussed in subsequent chapters.

Apart from resting metabolism, there are two other main ways that heat is produced, and the amount of heat varies in animals of different trophic levels. The first is the heat energy produced through the actual feeding process, rather than the food itself. This is termed the specific dynamic effect of food and is considerably higher in a carnivore diet (meat) than in a herbivore one (carbohydrate). This explains why large carnivores are often seen breathing heavily during and immediately after feeding. The second involves muscle activity, which also produces large amounts of heat. This is especially true in locomotion activities such as hunting, where the activity may increase heat production up to 10 times greater than that produced at the resting stage. Carnivores need to hunt for food and thus generate considerable heat from locomotive muscle action, particularly when compared to herbivores whose food is stationary in the environment and can be accessed at a leisurely pace. These interesting differences in the primary consumer level (herbivores) and secondary consumer level (carnivores) are found in both endotherms and ectotherms.

Each thermoregulatory mechanism has advantages and disadvantages, and one may not necessarily be considered better than the other. There are also quite a few animals that use a combination of these extremes, through behavioural and/or physiological means.

CHAPTER 8
Shape and size considerations

The size of animals has been briefly considered in chapter 7 in relation to endothermy and ectothermy. The principles of organism shape in terms of surface area to volume ratio, as described and demonstrated in chapter 2, apply as much to animals as they do to plants and are expanded on in this chapter. There are a variety of different shapes found in consumers.

The endothermic guineafowl is a ground bird that has a geometric shape which approaches that of a sphere, the shape which gives the best surface area to volume ratio. The ectothermic Cape cobra, on the other hand, is a snake that has a long, thin, cylindrical shape and thus needs to adapt to the challenges of a body form that has a high surface area to volume ratio.

Even if shape is favourable, body size, thermoregulatory mechanism, diet type and activity patterns still need to be considered, as they all obviously influence energy requirements. Therefore, considering only one or two of these criteria is not always helpful in gaining an understanding of energy and energy flow.

> ### BOX 5: THE BIG FIVE AND LITTLE FIVE
>
> The use of the tourist attraction term, the 'Big Five', is not very useful in terms of considering energy flow, as the term implies that these are the most important animals in the ecosystem worth seeing. The term is a hangover from the days of trophy hunting, where the Big Five were the most difficult and/or dangerous to hunt. They are all endothermic mammals, three being herbivores (elephant, rhinoceros and buffalo) and two being apex predators (lion and leopard). The 'Small Five', championed by some tourist operators, are more representative of savanna animal diversity and diet variation, but are also limited in relation to understanding energy transfer. There are two endotherms, a mammal (the insectivorous elephant shrew) and a bird (the seed-eating buffalo weaver). Three ectotherms are included: one reptile (the herbivorous leopard tortoise) and two insects (the predatory antlion and the rhino beetle, which feeds on fruit, nectar and decaying plant matter). The value of using either the Big Five, Small Five or both is limited in terms of understanding energy and thermodynamic processes in the ecosystem. This will become clear as the challenges of shape and size, coupled with thermoregulation requirements, are discussed for individual animal groups.

The basic metabolic rate is used as a comparative measure in considering energy needs in both ectotherms and endotherms. The basic metabolic rate has been found to be proportional to body mass (M) raised to the power of 0.75 ($M^{0.75}$). As body mass increases, the basic metabolic mass decreases. The metabolic mass is equivalent to and used as an alternative to the surface area to volume ratio. The relationship between the two measures is probably best illustrated by means of an example.

Why does an elephant require proportionately less food than the much smaller steenbok?

The elephant, a large herbivore with a mass of about 3300 kg, is compared with a similar shaped small herbivore, the steenbok, weighing only 11 kg.

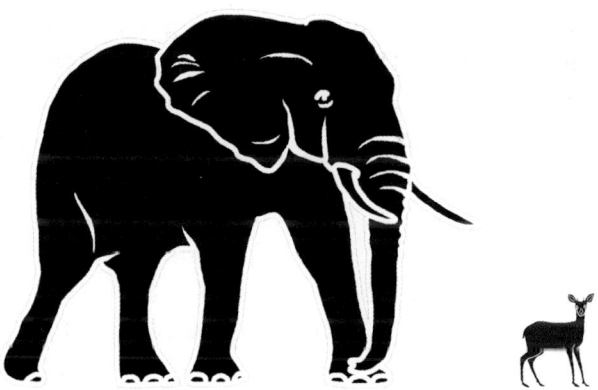

We already know from previous examples and worked calculations that the surface area to volume ratio of the large elephant must be considerably lower than that of the small steenbok. The elephant, having a larger surface area, exchanges energy with the environment more slowly than the steenbok and requires proportionately less food because of the differences in surface area to volume ratios. This can also be expressed in terms of their metabolic masses.

The elephant is some 300 times heavier than the steenbok and one might draw the simple conclusion that it would need 300 times more food to meet its energy requirements. The employment of metabolic mass indicates that this is not true. The metabolic mass of the elephant is 435 kg ($3300^{0.75}$) and that of the steenbok only 6 kg ($11^{0.75}$), so the elephant only requires 72.5 times (435/6) more food energy; approximately a quarter of what may be expected, if only body mass were considered. Another way of looking at the example is that the metabolic mass of the elephant is only 13% of its body mass, while for the steenbok it is 55%. Thus, a relatively large animal will have a reduced relative energy need than a smaller animal of the same basic shape.

The principles as outlined in this section are applied throughout the following chapters in this guide. The use and transfer of energy by animals in the savanna ecosystem is described for readers in the next three sections. In the main, metabolic mass is employed, but backed up with reference to surface area to volume ratios when considered necessary.

SECTION D
Primary Consumers: Herbivores

This section concentrates on those animals that feed mainly on graze and browse, as outlined in section B. The first chapter discusses herbivory in general, explaining why the main herbivores are endothermic mammals and ectothermic insects. Each of the final four chapters focuses on one of the major savanna herbivore guilds. These are the endothermic mammals and birds and the ectothermic reptiles and insects.

CHAPTER 9

Who are the herbivores?

The term herbivore refers to animals that feed on live plants and plant parts. Although there are various feeding approaches used by herbivores, the nutritional demands in terms of amino acids and energy producing sugars are similar. Some herbivores require their food to be in a concentrated form and feed on nutritionally rich plant food, such as fruits (e.g., many birds, baboons, monkeys and bats). Others that have relatively limited digestive systems, feed on sap and nectar (e.g., sunbirds), while many insects (e.g., caterpillars) and most of plains game (e.g., blue wildebeest, zebra and giraffe) can cope with large quantities of leaf roughage and modify its nutritional value through fermentation by micro-organisms. In this guide, the focus is on the leaf-eating herbivores, as only a relatively small proportion of the primary production of tree and grass plants is allocated to flowers, fruits and seeds. Reference to seeds will mainly be restricted to grass fruits (hard fruit). The fruit of trees, by comparison, is usually soft when ripe.

Herbaceous plants (small, soft-stemmed plants other than grass) are also not discussed in any detail, as, together with flowers, fruits and seeds, they probably account for less than 5% of the savanna primary production. However, it is worth mentioning that those herbivores (e.g., birds and bats) that do utilise these energy-rich plants and plant parts have the advantage of being able to assimilate a greater proportion of the plant material consumed (ingested) because of its higher quality. Assimilation refers to the part of ingested food that is processed by the digestive system and then absorbed by the animal to make new cells and tissues. That portion that is not absorbed is excreted from the animal body. Assimilation efficiency is thus the assimilated portion divided by the ingested amount. The assimilation efficiency for herbivores consuming the energy richer plant parts (fruit, nectar, etc.) is over 70%, while the assimilation efficiency of leaf-eating herbivores is usually only about 50%. This is because the challenging chemical composition of plant leaves makes them more difficult to digest efficiently.

BOX 6: ANIMAL GROUPS THAT CAN PROCESS LEAF TISSUE

In general, only three groups in the animal kingdom have adapted to breaking down tough green leaf tissue, which often contains greater than 30% of the structurally complex carbohydrate, cellulose. The three animal groups able to utilise this cellulose are the gastropod molluscs (snails), who possess a radula (horny tooth-like structure), insects with strong mandibles (mouth parts) and the herbivorous mammals that have well-developed molar teeth, providing a corrugated grinding surface. Attention in this book will focus on the mammalian and insect herbivores, as the gastropods probably only utilise a relatively small portion of leaf production.

Herbivores found in the savanna will now be considered by shape and size under the broad thermoregulatory headings of endotherms (mammals and birds) and ectotherms (reptiles and invertebrates, mainly insects). Little attention is devoted to considering juveniles, the examples focusing on mature animals. The young of endotherms are generally supplied with high-energy food by the mother, for example, milk supplied by mammals. Ectothermic young are for the most part on their own shortly after birth. One example of the differences in foraging behaviour, as related to the size differences in juveniles and adults, is discussed for predatory lizards in chapter 18.

CHAPTER 10

Endothermic herbivores: Mammals

Herbivorous mammals range in size from exceedingly small (e.g., rodents) to exceptionally large (e.g., elephants). The metabolic requirements, as we have seen, are different at these two extremes. Herbivores of different sizes thus have different dietary requirements and varied digestive systems to cope with their specific metabolic demands. Essentially two digestive systems are employed by herbivores. These are the foregut and hindgut systems, which are introduced by considering two common middle-sized herbivores which feed only on grass. Thereafter, the dietary and digestive requirements related to animal size and energy needs are discussed for selected herbivores. No attempt is made to consider all mammalian endotherms in savanna, only a few from different functional groups are described.

The zebra and blue wildebeest
Middle-sized grazing companions with different digestive systems

These two middle-sized herbivores are often seen grazing together. They are often referred to in terms of annual migrations in savannas, notably the Serengeti-Mara migration in East Africa. Historically, they were partners in local annual migrations even in southern Africa, but these movements are now severely curtailed by fences. These included migrations in and out of the Kruger National Park in South Africa, to and from the Okavango Delta in Botswana and around the fringes of Etosha Pan in Namibia.

There are several advantages gained from this communal living. Zebras have excellent eyesight and wildebeest have superior hearing, which enables them to cooperate in detecting possible predators more efficiently. Wildebeest are also said to have an excellent nose for rain, up to 25 km away, and can thus assist the zebra in moving to areas where fresh grazing would be available. Zebras prefer longer grass and open the grass sward for the wildebeest, who prefer the shorter fresh grass parts. This means there is cooperation, rather than competition, for food energy, even when feeding on the same species, and both have specific adaptations to cope with their preferences.

The zebra (about 260 kg) is larger than the wildebeest (about 160 kg) and they have metabolic masses of 65 kg and 45 kg, respectively. The zebra is about 1.6 times heavier than a wildebeest, but its metabolic mass is only about 1.4 times greater. The larger zebra thus needs proportionately less protein than the wildebeest to keep going (since it requires proportionately less energy) and is thus comfortable grazing on tall grass, whose cells might contain considerably greater quantities of cellulose (fibre) and relatively smaller amounts of protein than shorter grasses.

The zebra: The only mid-sized hindgut herbivore

The zebra satisfies its energy requirements by favouring quantity over quality, in terms of its forage intake, and is the only medium-sized open plains herbivore with a hindgut digestive system.

The digestive system of the zebra allows it to process far greater quantities of forage than its metabolic mass suggests. In fact, it may consume up to 4% of its body weight in food per day, whereas the wildebeest only consumes up to 2% of its body weight per day. The zebra, unlike most of the large savanna herbivores, has incisors in the upper and lower jaws enabling it to crop the fibrous grass quite easily. It then chews thoroughly, breaking the plant cell wall and releasing the contents. The cell contents, which includes proteins, are readily processed, and absorbed across the simple stomach and small intestine walls, but the cellulose released from the cell wall can only be digested when it reaches the caecum and large intestine (hindgut).

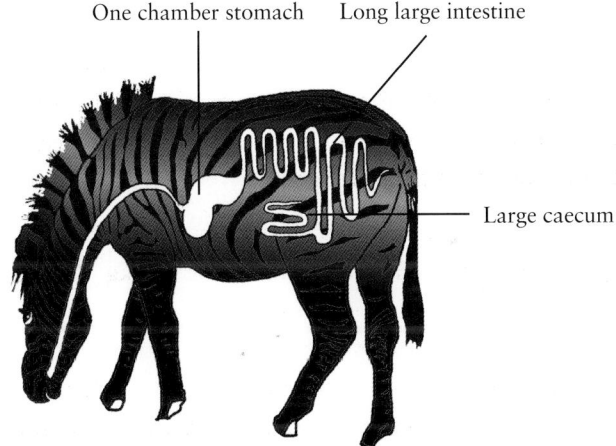

These posterior organs house microorganisms able to synthesise the enzyme cellulase. This enzyme breaks down cellulose and lignin (also a structural support compound of plant cell walls) and processes them into simpler compounds, which can then be transported across the large intestinal wall.

Digestion of cellulose thus takes place in the hindgut of the zebra.

They need to graze for about 15 hours a day to gain enough energy from food intake to support their activities. One seldom sees a thin zebra, partly because it can digest low-quality dry season coarse grass, which is of no use to its ruminant cousins. The zebra, in fact, eats like a horse!

The wildebeest: A mid-sized grazing ruminant

The wildebeest completes most of its digestion, including that of cellulose, in the foregut. Digestion takes place in a four-chambered stomach and larger food particles (the cud) are regurgitated for re-chewing (ruminating). The ruminant process requires the wildebeest to spend about six hours a day chewing the cud, in addition to about eight hours spent on grazing. The wildebeest has no incisors in the upper jaw and plucks the soft nutritional leaves off shorter grasses by manipulating them between the jaws. It does not need to breakdown the cell walls in the mouth as thoroughly as the zebra, partly because the cell walls will be chemically broken down in the stomach and partly because the grass parts not broken down sufficiently in the stomach, are returned to the mouth, in the form of the cud, for further chewing.

The ruminant's non-absorptive front stomach is separated into three parts – the rumen, reticulum and omasum – with a fourth part being the true stomach where digestion takes place. The food is initially retained in the first two parts (the rumen and reticulum). Microorganisms break down cellulose, and boluses of food that need to be further masticated are returned to the mouth. When food particles are small enough, they are passed on to the third chamber (the omasum) and finally the true stomach (abomasum), where the digestive process is completed and most of the product is absorbed across the walls of the small intestine and into the body.

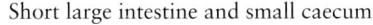
Endothermic herbivores: Mammals

Short large intestine and small caecum

Four-chambered stomach

BOX 7: WHAT ARE THE ADVANTAGES AND DISADVANTAGES OF THE RUMINANT AND HINDGUT DIGESTIVE SYSTEMS?

In the zebra the cellulose (a big part of potential energy) is still in the cell walls and is not digested until it reaches the caecum and large intestine in the hindgut. Nutrient uptake in this posterior part of the intestine is not as efficient as uptake from the small intestine, so the zebra loses some of the potential food energy in the faeces. The wildebeest, on the other hand, breaks down cellulose before it reaches the small intestine and, along with the digestion of all other cell contents, allows absorption of chemicals across the gut wall in the more effective small intestine. It might

> be thought of as a disadvantage that the microorganisms are so efficient in the rumen because they also use the products for their own population increase. However, large numbers of microorganisms themselves are passed onto the true stomach and the ruminant ultimately gets the nutrients that were produced earlier. In fact, most of the protein is used and the essential amino acids needed are manufactured from this source. In contrast, the hindgut fermenter must usually obtain all the essential amino acids from the plants it eats. Another advantage of the ruminant system is that the microorganisms in the foregut remove harmful chemical compounds, whereas, in the zebra, plant toxins are absorbed into the bloodstream and detoxification takes place in the liver.

When grazing, the zebra focuses on consuming quantity rather than quality. Rapid processing moves food through the system in about 35 hours, while in the wildebeest it could take 80 hours or longer. Although the zebra is not as efficient at getting energy from the important cellulose, it makes up for this by processing a large volume of food quickly and obtaining enough energy from the cell contents in a relatively short space of time. The wildebeest can extract the maximum amount of energy from the cellulose, but only if the grass has a relatively low fibre content. High fibre material found in coarse grass and stems takes a long time to break down into small particles and thus slows down flow through the system. As we shall see later, this could be a major challenge for ruminant herbivores that are much larger or much smaller than the wildebeest.

BOX 8: MIDDLE-SIZED RUMINANT GRAZERS

Like the wildebeest, most of the middle-sized (100–250 kg) common herbivores in the savanna are ruminants and primarily grazers, as opposed to being mixed feeders or browsers. These ruminant grazers include tsessebe (125 kg), hartebeest spp. (160 kg), waterbuck (237 kg), sable (210 kg), roan (250 kg), gemsbok (160 kg) and non-savanna herbivores such as the black wildebeest (150 kg), blesbok (80+ kg) and bontebok (90+ kg).

The different middle-sized savanna species use different grazing areas and/or different species or plant parts. Hartebeest are often found in the same areas as wildebeest and zebras, but do not compete directly with them, as the hartebeest seek out fresh grazing, particularly in recently burnt areas. The sable, roan and tsessebe prefer the longer grass areas, often associated with sour-veld, waterbuck concentrate on areas close to water and are non-selective grazers, while the gemsbok is a specialised selective grazer of the western arid savanna. There is thus virtually no competition amongst these mid-sized ruminant grazers of the savanna.

The upper limit for a purely grazing ruminant is about 250 kg. The larger the animal, the longer the passage time through the digestive system would be. If this becomes too long, about 100 hours, problems could arise. One suggestion is that, if the food is processed too slowly, methanogenic bacteria in the gut may start to attack it. Much of the energy is then captured by the bacteria themselves, rather than the herbivore, and vast numbers of the bacteria and protozoa may be lost in faeces rather than absorbed.

The buffalo, a bulk ruminant grazer weighing about 500 kg, is an exception to this limitation. It has a better ability to digest coarse grass than any other ruminant, but there is also evidence that it selects different grasses throughout the year, concentrating on the sweetest and thus highest in protein at any time. Buffalo herds cover vast distances at night searching for the best grazing sites. The better the graze quality, the less time is needed for food to pass through the rumen.

The kudu, at about 280 kg, is also a ruminant, but virtually an exclusive browser. Because browse is of higher quality and more easily digested than graze, the passage time through the digestive system is reduced. There are a few even larger ruminants, or foregut fermenters, that have overcome the passage-time challenge, either by concentrating on very high-quality leaves and/or having physiological and/or behavioural advantages that limit the passage time for digesta throughflow.

The giraffe, eland and hippopotamus
Very large foregut fermenters

The giraffe and eland, both ruminants, often weighing in at over 800 kg and on the large side for a ruminant, are either mixed feeders or browsers.

The giraffe, the tallest land mammal, has evolved special harvesting structures to ensure it feeds on the most nutritious tree leaves (browse) high up in the canopy. Because of its size, the giraffe has a relatively low metabolic mass (about 17% of its body mass). It thus obtains sufficient energy from consuming only 3.4% of its body weight in browse per day. It is also able to pluck the most nutritious leaves with the highest protein content from between the thorns by using its narrow muzzle and long tongue (up to 45 cm). It also has the advantage that the leaves are virtually out of reach of any other herbivore, so competition for quality and quantity is almost non-existent. The giraffe has densely packed surface papillae in its rumen, which increases the surface area for absorption by up to 30%. The high quality of its food source (browse) allows for easy breakdown into very small particles and thus reduces gut retention time, so that it does not approach the critical 100 hours. Digestion takes place rapidly and the giraffe ruminates contentedly.

The eland, weighing about 850 kg, is the largest antelope in the savanna and is unusual in that it is nomadic. The nomadic habit is a superb adaptation, enabling the eland to follow the rain and seek high-quality growth found in areas of unpredictable and localised rainfall patterns. The eland utilises the fresh green leaves of grasses in summer and switches to browsing in winter, focusing on high-quality foliage, such as mopane leaves, which are rich in protein. Thus, like the giraffe, the eland is a contented ruminant, despite its large size.

The hippopotamus, weighing about 1800 kg, would also seem to be too large to be a ruminant. It is unique, though, as it is a foregut fermenter, but not a true ruminant.

It only has three stomach chambers, and it does not chew the cud. Hippos, surprisingly for their size, are grazers and use their large lips to pluck leaves, mostly from short grasslands. Unlike giraffe and eland, they do not necessarily seek out the most nutritious leaves. They save a huge amount of energy by spending most of their time in water, which is not subject to the great temperature fluctuations found in terrestrial habitats. This more stable temperature regime means that they require less food than their body size would suggest. The hippopotamus may only need about 25 kg of graze per day, which is only about 1.5% of its body mass, much lower than most herbivores. This means that the relatively small quantity of relatively low-quality feed is processed in the large three-chambered foregut and the passage time through the digestive system will be much lower than might be expected for such a large herbivore.

The elephant and rhinoceros
Very large hindgut fermenters

The elephant, weighing 2500 kg or more, is a very large herbivore. It utilises graze and browse, while the white rhinoceros (1800 kg) is exclusively a grazer and the black rhinoceros (1600 kg) exclusively a browser.

Firstly, having a very large body with a shape that tends towards the spherical, these herbivores have relatively low metabolic masses, which are only about 15% of their body weights. They thus require proportionately less food energy than, for example, the zebra which is a hindgut fermenter but has a metabolic mass of about 25% of its body mass. The three large herbivores can get by with as little as 2% of their body mass in dry food per day. Although the actual percentage of food needed might be low, the absolute quantity that needs to be processed is large in comparison to the mid-sized herbivores. If they were ruminants, the digestion rate would be extremely slow and passage through the system would be much longer than 100 hours. They have adapted to this constraint by having a hindgut fermentation system that functions like that of the zebra. Selecting foliage with high nutritional value also speeds up the digestion process. The black rhinoceros makes exclusive use of highly nutritious browse, while the elephant, much like the eland, alternates between nutritious grass and browse and the white rhino is largely only a grazer, utilising the more nutritious young leaves of the shorter grasses.

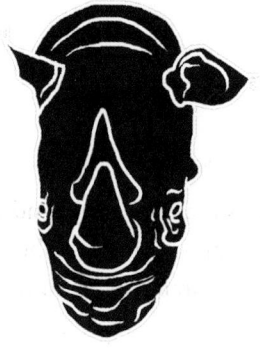 

Black rhinoceros head showing hooked lip used in browsing

White rhinoceros head showing the square lips used to crop grasses

Like the zebra, the white rhinoceros has teeth adapted to crushing grasses thoroughly before swallowing. Its digestive system is more efficient than that of the black rhinoceros in that it stores fibrous material for longer and is more efficient in digesting fibre. The white rhinoceros is restricted to areas receiving sufficient rainfall to support nutritious perennial grasses, whereas the elephant and black rhinoceros utilise the browse available in drier areas such as western Namibia. The elephant also feeds on tree bark and roots in the unfavourable winter season.

Therefore, the hindgut fermentation system is advantageous to these mega-herbivores, as it allows them to process large quantities of food rapidly. The challenges associated with a long digestion time are thus avoided, but digestion is less efficient than that of the ruminant system. Food, therefore, needs to be consumed in large quantities, as much of it will be excreted in the form of poorly digested dung.

There are also changes in diet type and digestive systems of the herbivores that are smaller than the middle-sized (100–250 kg) grazing ruminants previously discussed.

The impala, springbuck, southern reedbuck and nyala
Smaller ruminant mixed feeders

Mammalian herbivores that weigh less than the lower end (100 kg) of the middle-sized grazing ruminants, down to about 35 kg, fall into this group. The energy requirements of ruminants of this size, dictate that they must become mixed feeders. The effectiveness of ruminant digestion is related to body size. A middle-sized grazing ruminant, such as the blue wildebeest, has a relatively voluminous rumen to process food in terms of its metabolic mass and therefore has more space to process food than a smaller one. As body size decreases and relative metabolic mass increases, the energy demands of the small herbivores are such that they must eat food of higher quality.

Springbok

Furthermore, they need to extract as much energy as possible from this as their relatively small rumens can only support a small volume of food. Two of the herbivores in this group, the impala and springbuck, will be discussed.

Impala

The impala, weighing about 45 kg, and the springbuck, about 38 kg, are relatively small ruminant mixed feeders. They utilise browse and graze in about equal proportion.

The impala has a relatively high metabolic mass of 17 kg and the springbok, with a metabolic mass of 15 kg, must include the more highly nutritious browse in their diet which can be processed more speedily through their small rumens. They do not compete, however, because they do not occur together. The springbuck is typically found in arid areas, while the impala frequents the moister savanna.

The nyala and southern reedbuck are also mixed feeders but are not found in large herds, typical of the impala and springbuck. They both prefer thicker vegetation or tall grass habitats where they can conceal themselves. These areas are usually close to a water source.

The bushbuck, duiker, steenbok, klipspringer and dik-dik
Very small browsing ruminants

Most of the herbivores in this group fall in the 5 kg to 35 kg range. Their metabolic masses are much higher than the previous size group and they need to feed almost exclusively on high quality browse, to compensate for their relatively small rumen volumes. They have special adaptations to meet exceptionally small ruminants. They have relatively wider mouths, longer tongues and teeth better adapted to macerating than grazing ruminants. In addition, their digestive systems have modifications. They have relatively large salivary glands, a larger opening between the reticulum and omasum and larger true stomachs, intestines and caeca, than their larger grazing and mixed feeder ruminant cousins. All the above adaptations assist in the efficient harvesting and digestion of browse and ensure that the food flows rapidly through the system.

These small ruminants are very sedentary, as they need to save as much energy as possible. They are usually found in pairs, rather than large herds, thus ensuring that the available forage is available to only a few individuals and the best parts can be sourced. The smallest ruminant in the southern African savanna is the dik-dik, but it is restricted to Namibia and is only about 5 kg in weight. To assist in ensuring rapid digestion and quick through flow, it includes high-quality fruit in its diet.

Herbivores smaller than the dik-dik must, out of necessity, be hindgut fermenters. There are, however, also two herbivores weighing in the range between 25 kg and 80 kg who are hindgut fermenters. These are the warthog and porcupine.

The warthog and porcupine
Unusual smaller hindgut fermenters

The adult warthog weighs about 80 kg on average and is the only member of the pig family in the savanna. It has very short legs in comparison to ungulates of a similar size and is thus not as mobile in terms of escaping from predators. It also has a largely naked body and is therefore often seen covered in mud to control excessive temperatures. It is predominantly a grazer, feeding on nutritious grass, but will also utilise high-energy plant parts such as tubers and roots. As a hindgut fermenter, it processes the food intake rapidly and is not limited by any constraints that might be associated with a ruminant digestive system.

The largest rodent in southern Africa, the porcupine, weighs up to 25 kg and is also a hindgut fermenter. It has a vegetarian diet ranging from underground food such as bulbs, tubers and roots, through gnawing at tree bark, to feeding on high-quality plant berries. Its teeth are adapted to breaking down cell walls and it has a large caecum where fermentation takes place. The digestive process is very efficient and extracts up to 85% of the protein from the food source.

The hares, small rodents and the dassie
Exceedingly small hindgut fermenters

Hares and rodents usually weigh less than 5 kg and this means that their metabolic mass would be too high for them to be ruminants. The long retention time that would be needed if they were ruminants means they would probably not be able to eat enough per day to survive. These small herbivores are thus hindgut fermenters and food ferments in the caecum. Many of them are not able to absorb much of the initial digestive product, but eat the faeces to recycle nutrients, a process termed coprophagy. Chewing the faecal cud!

The dassie (rock hyrax), at about 4.5 kg, looks like a rodent but was once thought to be more closely related to the elephant. It is thus not surprising that it has a pair of long, tusk-like incisors, reminiscent of the elephant. It uses these to scrape away at the bark of trees and receives moisture and nutrients from the inner bark. It, however, mostly uses its cheek teeth for cropping foliage. The foliage passes through the stomach to a large sac-like structure where fermentation bacteria are present to digest cellulose. This sack joins onto the caecum, which has two horn-like structures. These are involved in fatty acid production, which is an additional energy source. Digestion is, however, awfully slow compared to other small endotherms, and it does not adopt coprophagy.

Most of the smallest rodents and mice, weighing less than 0.5 kg, are also hindgut fermenters. Because they are so small, they must feed on high-quality plant food such as seeds and fruits. In fact, many are omnivorous and need to feed on high-quality animal tissue as well, including many invertebrates in their diet.

Thus, at the smaller herbivore size, their energy demands are such that they must be hindgut fermenters, must utilise a high-quality diet (browse, fruit, seeds, etc.) and some even require specialised digestive systems.

Summary of diet selection and diet type in savanna herbivorous mammals

The digestion system and size of the mammalian herbivores largely dictate their place in the ecosystem and this summary considers the various groupings by digestive system. Examples of how this information can be used to get an idea of the mix of species and their numbers, which can be sustainably run on a defined savanna ecosystem, are included in the appendix.

Hindgut fermenters

Hindgut fermentation is prevalent in large, greater than 1000 kg, (e.g., elephant, rhinos) and small, less than 5 kg (e.g., hares, rodents) herbivores. This is a result of body size and relative energy requirements. Large herbivores, despite relatively low energy requirements, still need to consume vast quantities of plant matter and need to have a rapid passage of digesta. The extremely high energy demands of the extremely small endothermic herbivores dictate that comparatively large quantities of forage are required. A ruminant digestive system could not cope with the quantity. Quality of forage is also important, so most of these hindgut fermenters will be browsers rather than grazers, as browse is of a higher nutritional quality.

Elephants are mixed feeders, switching from graze to browse in winter. Other mixed feeders are the porcupine and warthog. The black rhinoceros is a browser, while the white rhinoceros is exclusively a grazer. The zebra is the only middle-sized hindgut fermenter that is a dedicated grazer.

Foregut fermenters: Ruminants

There are no exceedingly small or very large ruminants. Most ruminants are grazers in the middle-sized category, 100–250 kg. On either side of this range, ruminants tend to be mixed feeders and at small (6–16 kg) and large (about 800 kg) sizes require high-quality forage of browse. The kudu, at 250 kg, is, however, an exception, as it is a ruminant browser, but not very selective, and may feed on herbs and even fresh grass at times.

Foregut fermenter: Non-ruminant

The hippopotamus is unusual in that it is neither a hindgut fermenter nor a ruminant. It is an exceptionally large (greater than 1000 kg) grazer, but needs relatively little forage for its size, as it spends most of the day in water. It feeds at night, selecting short nutritious grasses. Unlike ruminants, it has a three-, not four-, chambered stomach and does not chew the cud.

BOX 9: FOOD FOR THOUGHT

It is interesting that the five largest herbivores (1000+ kg) each have distinct combinations of diet and digestive systems. The hippo is a grazing, non-ruminant foregut fermenter, the giraffe is a browsing ruminant, the black rhino a browsing hindgut fermenter, the white rhino a grazing hindgut fermenter and the largest, the elephant, is a mixed-feeding hindgut fermenter. This contrasts with the other end of the spectrum (herbivores weighing less than 5 kg), where the species richness is higher, and these many small herbivorous mammals are exclusively browsing hindgut fermenters.

CHAPTER 11

Endothermic herbivores: Birds

Why high-quality plant food is essential

The ability to fly is an obvious advantage to birds, but it comes at a considerable cost in terms of energy consumption. The muscles required for flight may account for 20% of the body mass and the power output per unit mass of these muscles can be up to 20 times more than that of most mammalian muscles. Flight can thus only be carried out efficiently by small bodies and the wing load is important in this regard. Wing load is calculated by dividing the weight of the bird by the surface area of the wings. Larger birds would require proportionally larger wings and thus, at a certain size, flying becomes impossible. The largest flying bird in the bushveld is the omnivorous kori bustard, which weighs about 10 kg.

Birds need an efficient system to transport oxygen to the flight muscles. They have relatively large hearts, high rates of blood flow and a unique lung system that efficiently maximises gaseous exchange. These adaptations also assist in getting rid of heat and carbon dioxide produced by high levels of muscular action used in flight. Other adaptations associated with flight are that birds have hollow, light bones and even the skull is light, several bones are fused and teeth are absent.

Most of the herbivorous birds are on the small side. They thus have a high energy requirement, like the very small mammals mentioned in chapter 10. Most of their assimilated energy is used to support the muscles during flight, and so less than 1% is available for growth and reproduction. Birds, therefore, need to consume high-quality plant food, which is digested in a hindgut fermentation manner. The function of teeth has been taken over by the muscular gizzard, which forms part of the digestive system. The diet not only has a high energy content, but it is easily digestible. Most birds digest much more rapidly than herbivorous mammals; as low as 5 minutes have been reported for orioles, 45 minutes for waxbills and 6 hours for the flightless ostrich. The latter is slower because the ostrich consumes a fair amount of leaf matter. The cellulose must be fermented in the paired caeca, slowing down the digestion and throughflow rate.

It is thus not surprising that very few (only about 3% of all birds) specialise in eating leaves (folivores), which have relatively poor energy content. For example, the Egyptian goose, common in areas close to water in savanna areas, feeds quite heavily on plant leaves.

BOX 10: ENERGY REQUIREMENTS OF A DOVE AND A HUMAN

Another way of considering the energy requirements of birds is to compare a common small bird with a relatively small human being. A common phrase used when I was young was: 'You eat like a bird', implying consuming little. How misleading is this? To answer this question a small dove is compared with a human being and comparative data is shown in Table 11.1.

Table 11.1: Comparative food energy requirements of a small dove and a human being

	Dove	Human
Weight	0.113 kg	70 kg
Daily food intake	0.017 kg (15% of weight)	1.75 kg (2.5% of weight)
Calories per day	70	2000
Calories per kg	619	29

The small dove consumes a much higher percentage of its weight in food and uses many more calories per kg than the human being weighing 70 kg. If this human 'ate like a bird', he/she would need to consume 10.5 kg (15% of weight) per day and it would not be long before the person became obese.

Even though birds generally have a body shape approaching that of a sphere, they still have a relatively high surface area to body mass ratio, because of their small size. This, coupled with the high energy requirement of flight, requires a constant supply of high-quality food, such as fruit and seeds. Because these food sources make up a comparatively small component of primary production, birds make a relatively low contribution to overall savanna biomass. All birds (herbivores and carnivores) probably only have a biomass in the moist savannas of about 0.05 tonnes/km² when compared to the 0.25 tonnes/km² of just one mammalian herbivore, the abundant impala.

One unusual advantage associated with flight is that most birds live up to four times longer than similar-sized mammals. One suggestion for this is that flying reduces the predation on birds and this contributes to a longer life.

Examples of birds that feed on high-energy plant parts in the savanna

Seedeaters

Waxbills, finches, canaries, manikins and francolins are seedeaters. These birds typically have stout and conical bills which equip the birds for crushing and de-husking hard seeds. Most have a relatively large crop for storage and a muscular gizzard for further mastication of seeds.

Fruiteaters

Mouse birds, barbets, bulbuls, starlings, parrots, louries and pigeons are examples of fruiteaters. The beak shape is less easy to define for this feeding group, as there are a variety of shapes and sizes. For example, parrots have strong, hooked beaks and louries have strong, deeply curved bills.

Nectar feeders

Sunbirds have long, slender, decurved bills for probing flowers and have a long tongue to assist in sipping up nectar.

Mixed feeders

Hornbills (fruit and seed), red-billed quelea (seeds and insects) and orioles (fruit, nectar, berries and insects) are examples of mixed feeders. Even some of the birds listed above as specialists will also feed on other food types. For example, the barbets which concentrate on fruit may also feed on insects at times. Most of the larger ground birds, such as guineafowl, francolin and bustards, are also omnivorous.

In exchange for the energy expended metabolising energy-rich fruit, nectar and seeds, the plants gain benefits such as pollination and seed dispersal from the birds that consume these foods.

Although the total biomass of herbivorous birds is relatively low, the population numbers of some bird species are remarkably high. A mixed feeder, the red-billed quelea, is often observed in huge flocks and aspects of its ecology are discussed below. A second detailed example considers the energy used in hovering and migrating birds, with the sunbird used as the hovering example. Finally, more detail is given on energy use by the largest non-flying bird, the ostrich.

The red-billed quelea: Punching above its weight

This species can be found in colonies of up to 5 million adults and a similar number of chicks in the breeding season. Such a colony may consume as much as 13 000 kg of insects and 1 million kg of grass seeds in the breeding cycle. A single bird, weighing about 15 g and only 5 cm in length, may consume 25 g of grass seeds per day. In 2010 it was estimated in the Kruger National Park that there were up to 33 million of these in the breeding season and they could consume up to 825 000 kg of seeds per day, not an insignificant share of the non-leaf portion of plant production. It must be remembered, however, that despite this species being the most abundant bird species on the planet, constituting about 50% of the biomass of all bird species in a savanna area, the actual population biomass is very low. Even in the breeding season, the biomass might only be about 0.025 tonnes/km^2. This is still low compared with the extremely high values for herbivorous invertebrates in the savanna.

It is not surprising that the flocks may travel up to 50 km a day between roosting sites and feeding grounds. These feeding grounds are usually associated with annual grass species that grow quickly and set seed within a few weeks after a rainfall event. The birds thus migrate to favourable feeding sites, and this alone is expensive, as they must build up their fat reserves to undertake long flights. Fat stores about twice the energy per gram than either proteins or carbohydrates. Fat reserves may be greater than 50% of a migrant bird's weight and is constantly converted to fuel for muscles during flight. There is no doubt, then, that they are a crucial part of ecosystem function and must disturb local sites during their movements. Presumably, many intact seeds pass rapidly through the digestive system and thus assist in spreading seeds to many localised areas.

Hovering and migration: Energy sapping activities

Sunbirds usually perch on flowers to feed on their nectar, but they may also hover. This, of course, is an energy-sapping activity, consuming up to 3 kilojoules per hour. Although sunbirds are tiny, usually weighing less than 15 g, they may store a few grams of fat, which can supply enough energy to feed on the wing for a few hours. Although nectar is a relatively rich energy source, its digestion is not without challenges. Nectar utilised by sunbirds is often dilute, containing about 20% sugars and 80% water. The challenge for the bird is getting rid of vast quantities of consumed water, which could weigh more than the bird itself. Sunbirds use considerable energy ensuring that excess water is removed from the body. As the protein content of nectar is very low, sunbirds need to feed on insects to ensure sufficient intake of protein needed to increase muscle mass.

It is also important to save energy at times. Some sunbirds save energy by entering a state of torpor when there is a low supply of food or during cold nights. Torpor is essentially a period of inactivity accompanied by a reduction in regulated body temperature. Nightjars, swifts and mouse birds may also use this energy-saving mechanism.

Although hovering is very demanding in terms of energy requirement, the cost of migratory flight is even more so. Migration can be defined as the regular seasonal movement of populations of birds from one area to another and back again. Some sunbirds will travel considerable distances at different times of the year to find their favourite plant food. About 200 bird species migrate to the savanna in the summer months, some from thousands of kilometres away and others from closer locations. Many migratory birds are high-quality feeders and time their arrival to coincide with the maximum availability of flowers, nectar, fruit and seeds. They leave again when they have built up enough fat reserves to cater for the energetically expensive return migration.

The ostrich: Large and flightless

The ostrich, commonly seen in savanna areas, is the largest and heaviest bird on earth. Males can stand as high as 2.5 m and weigh about 150 kg. They are flightless, have sturdy legs and strong leg muscles to support running. They reach about 70 km per hour at top speed but are also able to sustain running at lower speeds for half an hour or more. They are gregarious, live in flocks and eat whatever they can swallow. Like most birds, the ostrich has a gizzard that is particularly large, with powerful muscles that help break down all types of food. Hard objects, such as stones, are swallowed and assist with the grinding process in the gizzard. The ostrich gets most of its water from the plants it eats and avoids dehydration partly because its body shape is almost spherical. Ostriches can have a devastating effect on local grazing lands if they remain in one area. This is not often seen in natural environments, but the effect can be seen in many camps where ostriches are farmed commercially around southern Africa.

CHAPTER 12
Ectothermic herbivores: Reptiles

Why are there so few reptilian herbivores?

Like birds, very few reptiles are leaf-eating herbivores, a consequence of body size, body shape and feeding processes. Firstly, some 70% of lizards weigh less than 10 g and this is one of the reasons why these vertebrates are mainly ectotherms and need a high-energy, more nutritious, carnivorous diet. Secondly, their tooth structure and digestive anatomy would indicate that a leaf-eating herbivorous diet would not be efficient enough to meet the energy requirements of such small vertebrate body sizes. In addition, most lizards are dorsoventrally or laterally flattened, which increases the surface area to volume ratio and, therefore, their energy requirements. Thus, most lizards are carnivorous.

Snakes have long and relatively thin bodies, a shape that increases surface area to volume ratio substantially. It is thus not surprising that all snakes are carnivorous and essentially fit in the trophic structure at least one level above that of the primary consumers (herbivore level), in the secondary consumer (carnivore) level.

The small body sizes and/or specialised body shapes enable many of the reptilian ectotherms to occupy ecological niches that are not available to endotherms, even the small ones. These niches, however, mainly involve a carnivorous diet and are discussed in the chapters dealing with the secondary and higher trophic levels. Two of the very few plant-eating reptiles are discussed below.

The giant plated lizard: An omnivorous lizard

The giant plated lizard (very seldom seen) is a rare example of an omnivorous lizard. With a length of up to 70 cm, it is the second largest lizard in southern Africa after the carnivorous monitor lizard. It feeds on plants and small animal prey and, like its giant cousin the crocodile, stores fat in its tail. It is possible that the giant lizard has a less demanding energy requirement than other lizards, because its body is so much larger; it can thus utilise plant material when it needs to.

The tortoise: An herbivorous reptile

The tortoise, commonly seen on the roadside, is a relatively large, slow-moving ectotherm, weighing 20 kg or more. Like some of the smaller endothermic herbivores, it is a hindgut fermenter. Tortoises have serrated horny lips with which they tear off vegetation parts, including grass, flowers, fruits, young leaves of annual plants and succulents. The plant parts are not crushed in the manner adopted by mammalian hindgut fermenters; the small plant parts ingested are very gradually broken down in the stomach and small intestine. Metabolism, therefore, takes place slowly. The tortoise's hindgut system is, however, very efficient in allowing for moisture to be removed from the various plant parts and they therefore seldom need to drink water. They even have a special water storage sack called a bursa, which allows them to store water for dry times. The slow metabolism and efficient water retention system allow the tortoise to conserve energy, and they can live for 50 years or more.

CHAPTER 13

Ectothermic herbivores: Invertebrates

Invertebrate herbivores: Much more important than one may think

The invertebrates play a far more important role as primary consumers than is often thought. Many people will know of the 'damage' caused by invertebrates consuming garden plants. Most ecosystem accounts, however, focus on their role in the decomposition process, rather than their role in herbivory. Decomposition, of course, refers to feeding mainly on dead material, while herbivory is the feeding on live plants and plant parts above and below the soil surface. Nematodes, for example, feed on the live roots below the soil surface, while most of the herbivorous arthropods feed above ground.

Arthropods are invertebrate animals with an exoskeleton, segmented body and paired jointed appendages. The arthropods form the largest group of animals, and the insects, in terms of number of species and number of individuals, are the largest group within the arthropods. Non-insect arthropods, including scorpions, spiders, ticks, mites, centipedes and millipedes, are mostly carnivorous. The discussion in terms of invertebrate herbivory will thus focus on insects.

Insects: The most diverse savanna herbivores

Insects, on average, weigh only about 3 mg and are therefore ectotherms. Because of their enormous diversity, they have managed to occupy just about every feeding niche one could think of. There are herbivores, secondary and tertiary consumers (carnivores), omnivores and, of course, numerous decomposers. Insects change shape during their lifetime, passing through up to four stages of development. For example, the butterfly passes through a larval stage in the form of a caterpillar. The adult stage is usually the shortest duration of all. An insect has three pairs of legs and the body, divided into three parts, has a hard outer casing. They have strong mandibles and a three-chambered complete digestive tract, including foregut, midgut and hindgut. Depending on what their main food is, they have the appropriate enzymes to break down the food into smaller particles for absorption. In addition to the complete digestive system, they have a very efficient tracheal tube system for transport of oxygen to body sites and removal of carbon dioxide from the body. The energy cost of flight is also much less in insects than in birds. Although the metabolic costs of actual lift are similar, insects produce twice the power per muscle mass. All the above attributes enable many insects to be leaf-eating herbivores, despite their small size.

Nymph or larval stages of many insects have a major impact as folivores and are superb herbivores, putting the larger vertebrate endotherms, the mammals and birds, to shame. Being ectotherms, they are much more efficient at converting assimilated energy from plant parts into growth than the small endotherms. Some may reach a production efficiency of up to 75%, compared to about 10% for mammals and 1% for birds.

BOX 11: INSECT BIOMASS GREATER THAN ALL OTHER ANIMAL GROUPS

Insects found in the canopies of trees in savanna may have a total biomass of 2.5 tonnes/km^2, which is roughly the same biomass per km^2 as all the mammals. If one added the ground insects, such as termites and grasshoppers, the biomass of insects could total about 7 tonnes/km^2. Insects are thus very efficient at converting plant intake into body mass and new generation insect biomass.

Only about 30% of the plant production in savanna is consumed by all the herbivores combined and insects account for up to 15% of this consumption. This may seem like rather a small percentage, but when the advantages of insect production efficiency are factored in, it is evident that insects convert a high proportion of primary production energy into biomass. The remaining 70% of the plant production not utilised is passed on as decaying and dead matter to decomposers and these are mainly insects. The actual amount of primary production will vary between dry and wet years and up to 50% of the production may be removed by the action of fire in any particular year. No detail on the types and frequencies of fire is considered in this guide.

There are many specialised plant-eating insects, including those that feed on the high-energy plant parts such as nectar (butterflies, hawkmoths), flowers (blister and flower beetles), hard seed (seed bugs) and pollen (honey and solitary bees). These all contribute to ecosystem services such as pollination and seed dispersal, even if the high-energy plant parts they use only represent a fraction of the primary production. One can hardly imagine the role that all these play in the ecosystem. About half the known insect species are herbivores, which is not surprising because insects and plants made their appearances on earth at the same time, some 500 million years ago.

Another way of looking at this immense harvesting of energy from plants is that these small ectotherms have re-packaged energy and thus make it available to other animals who are their predators and do not feed on plants. These would include the spiders, scorpions and carnivorous insects, the lizards, birds and even small mammalian carnivores. This aspect is discussed further in subsequent chapters.

Three leaf-eating insect examples are presented below, to emphasise the diversity of feeding forms.

The emperor moth: One of about 10 000 local species of moth and butterfly

The emperor moth is one of the larger insects with a wingspan of some 180 mm. One would be lucky to catch sight of this moth, however, as it only lives in this adult form for 4 to 5 days at the beginning of summer. The adults do not feed at all but spend their short existence searching for mates. Eggs are laid on a mopane leaf and they hatch during the summer. The larval stage, the mopane worm (caterpillar), crunches away on the leaf on which it emerged. The mopane leaf is the

favoured source, as it has a remarkably high protein content (16%) when compared to other trees and the caterpillar seems to be unaffected by tannin levels. The caterpillars feed so voraciously that one can even hear the crunching if close by and they can denude local mopane tree patches of all their leaves. Fortunately, this stage of their life cycle lasts only a few months, so not too many trees are defoliated.

As it grows larger, the mopane worm goes through four moulting stages, then drops to the ground, buries itself, and the pupa stage begins. This stage lasts about seven months, through the winter, until the insect emerges in its adult form at the beginning of summer. This is a good example of an ectothermic animal which changes its shape and body size and thus energy requirements dramatically during its life cycle.

The mopane worm is harvested as a food source by human communities. It is not unusual for harvesting to yield over 20 kg/ha, excluding what has already been consumed by the natural biota. Predation on the various stages of the life cycle is high, providing food for at least 20 insect species, four reptile species, 34 bird species and 10 mammal species, including bat-eared foxes and jackals.

Cicadas: About 140 local species

Another common insect which is more likely to be heard than seen is the cicada. The adults, which only live for a few weeks, can be heard calling for mates early in the summer months. The males sing (if that is the correct term for the din they create!) by flexing their tymbals, the drum-like organs found in the abdomen. The adults feed by sucking juices out of leaves. After mating, the female deposits her eggs on a small branch, in a groove from which plant juices can ooze. When the young nymph emerges from the egg, it sucks up the energy-rich juices in the groove and then falls to the ground where it buries itself, sometimes for years, living off root sap. Damage to the tree is generally minimal. When the nymph re-emerges, it crawls up a grass or tree stem, metamorphoses into the adult form and the cycle starts again.

Katydids and grasshoppers

Other leaf consumers include the katydids (about 100 local species) and their close relatives, the grasshoppers (about 600 local species) and locusts. The katydids have excellent camouflage and resemble, in body form, the very leaves on which they feed. All grasshoppers and locusts feed exclusively on plants, particularly leaves. They are essentially the same insects, called grasshoppers when in the solitary phase and locusts in the swarm phase. The devastation caused by locusts is usually restricted to the drier savanna areas, but even in the moist areas, they can cause local devastation if conditions are right for swarming. The grasshopper adults live for only about two months, but the nymphs are also responsible for herbivory. An adult weighing 300 mg can eat between 30 and 100 mg of dry matter per day. Grasshoppers, sometimes in swarms of millions, can remove up to 15% of the annual primary grass production in moist broad-leafed savanna areas. This is because grasshopper biomass can be up to 75% of total insect biomass in broad-leafed savanna and over 90% in fine-leafed savanna. Density can be as high as 10 individuals per m^2, which equates to a biomass of about 2.3 tonnes/km^2. Grasshoppers can account for over 40% of all herbivore consumption; the consumption on fine-leaved savanna is over 40 tonnes/km^2/year and in broad-leaved savanna about 13 tonnes/km^2/year – not insignificant amounts.

SECTION E
Animals That Feed on Animals: Secondary Consumers and Beyond

In basic descriptions of the secondary consumer level, attention is often focused primarily on the carnivorous (meat eating) predators. However, many predators will consume insects and other invertebrates whose bodies are not primarily composed of meat in the traditional sense but are nevertheless composed of animal tissue. These are included in this section. Furthermore, the word predator is usually used in a situation where one animal hunts, kills and eats another animal; however, scavengers are also included as consumers in this section, although they do not hunt or kill the animals on which they feed.

Only about 10% of the net primary production energy captured by herbivores is available for use by secondary consumers. Thus, all the animals at the secondary consumer (predator) level need to be highly efficient in utilising the relatively small amount of available energy.

The first chapter in this section (chapter 14) considers all the types of secondary consumers feeding as predators. The next three chapters consider the roles of the endothermic predators and the final two are concerned with the ectothermic predators.

CHAPTER 14
The predators of animals

Many animals that eat other animals may also, at certain times, feed on plant matter (primary producers) and are thus omnivores. Omnivores have higher metabolic rates than herbivores and would normally prefer to eat animal matter and thus only feed on plant material, which has less protein content, when the high-energy animal food source is scarce. This section, however, does not consider omnivores in any detail. The focus is on the animal-eating predators. The energy requirement of a vertebrate secondary consumer is calculated using metabolic mass, in the same way as described in the previous chapters for primary consumers.

Not only are mammals and birds (endotherms) important predators but, as previously mentioned, most of the vertebrate ectothermic reptiles (crocodiles, lizards and snakes) are also predators and they themselves may be preyed on by larger predators. Many invertebrates, including insects, are also predators, feeding on other invertebrates and/or birds or even small mammals, before being preyed on themselves. Many of the predatory ectotherms, such as lizards, snakes and invertebrates, feed on prey that occupy spaces not available to the larger endothermic predators. These ectotherm predators, which may have net production efficiencies (growth and reproduction) of over 75%, thus repack their preyed-upon energy into their own or their offspring biomass, which is then available to many other predators. This is significant when it is considered that birds and mammals generally have production efficiencies below 10% and thus contribute proportionately less assimilated energy per individual to the next level in the food chain. The small lizards and invertebrate predators are thus able to efficiently utilise the smaller portion of energy passed on from the plants to the ectothermic herbivores, mainly insects. These small predators are thus able to reach high population numbers, which translate into a significant amount of biomass. Even the larger ectotherms may attain high densities and biomass. For example, in an area close to a waterbody, pythons, crocodiles and monitor lizards can surpass mammalian carnivores both in numbers and biomass by several orders of magnitude. No attempt is made to place the predators into more than one trophic level; predation on each animal group is discussed separately when needed. There are many predators that operate at more than one trophic level and trying to allocate any one of these to a particular level becomes quite difficult, especially in terms of explaining energy transfer.

The apex mammalian carnivores, such as the lion, spend a considerable amount of time resting, conserving energy for actual hunting. Many of the ectothermic predators (crocodiles, lizards, snakes, and some arthropods) adopt an energy-saving, sit-and-wait strategy (ambush), while others forage over a wide range, which is more energetically demanding. Smaller perching insectivorous birds also adopt one or other of these two strategies. The shapes and sizes of the animals vary according to which of the two strategies they adopt, and these are discussed under each group.

BOX 12: CHALLENGES FOR ORGANISMS THAT EAT ANIMALS

There are numerous challenges, but also advantages, in being an animal that eats other animals rather than plants. Herbivores have the advantage in that their food source is relatively plentiful and stationary, whereas predators spend considerable time and energy in the search, pursuit and capture of their relatively rare and mobile food sources. However, animal tissue is predominantly protein, which lacks the constraints associated with digesting the cellulose present in plant cell walls. Predators thus also have relatively high assimilation efficiencies, but production efficiencies are very low.

Predators consume food that is low in dry matter and high in protein, which can be digested rapidly to produce energy. Little chewing is needed, and a simple digestive system allows for rapid absorption and throughflow, with a relatively small amount of non-digestible content lost in faeces. Although the intestines of carnivores are about six times longer than their body length, this is much shorter in comparison to herbivores, whose intestines need to be some 24 times their body length to facilitate the digestion of their carbohydrate diet, particularly cellulose.

Proteins and carbohydrates contain the same amount of energy, about 4 calories per gram. The energy cost of digesting protein is, however, greater, at about 25% of the total daily energy costs, than that for carbohydrates, which is in the range of 5–10%. The latter may be higher for ruminants, who must expend considerable energy metabolising the complex cellulose carbohydrate. In the end, ruminants and other herbivores are well adapted to a high carbohydrate diet and obtain sufficient protein from their forage and absorption of microorganisms. The animal eaters have a diet which supplies sufficient protein, and produce carbohydrates (e.g., glucose) when needed, from amino acids that have been absorbed across the gut wall, and fat. Despite the pros and cons of the varied energy costs and energy supply of the two dietary types, both carnivores and herbivores are successful in their respective trophic levels and are composed of species representing all shapes and sizes.

CHAPTER 15
Endothermic predators: Mammals

The larger predators: Greater than 20 kg

This group includes all the carnivorous predators larger than the caracal, the male of which typically weighs in at about 19 kg. The spotted hyena, wild dog, cheetah, leopard and, of course, the largest predator, the lion (the male weighing about 250 kg), are typical meat-eating members of this dietary group. The aardvark, which weighs in at about 60 kg when mature, also falls into this weight group but it is insectivorous, its diet consisting mostly of termites.

The meat-eating predators weighing more than 20 kg regularly feed on prey that is considerably larger than their own bodies. They expend considerable energy while searching for, catching and finally subduing their very mobile prey. Their success rate is often less than 50%; thus, for at least every second hunt, energy is wasted on unsuccessful pursuit.

A small prey specialist, such as the caracal, would spend twice as much energy in the pursuit and eventual catching of its target if it switched to preying on animals that were much larger than itself. This would not be desirable, however, as the larger prey would not compensate for the two-fold increase in energy use by the caracal, especially as there could be intense pressure on the small caracal to vacate the kill by larger carnivores and scavengers. A 20 kg caracal, although only one-ninth of the weight of a 180 kg lion, has about one-fifth of the metabolic mass of the lion. The larger predator would thus need proportionately less energy from food than one weighing 20 kg or less. A lion of the above weight would require about 5 kg of meat per day, which is 2.8% of its body weight. Although the much smaller caracal only requires about 1 kg per day, this equates to about 5% of its body weight.

The larger carnivores must, nevertheless, adopt energy saving tactics to counteract the energy expenditure needed in pursuit of large prey. The lion, for example, appears to be very lethargic, resting for up to 20 hours per day, but this is necessary for recovery after energy-sapping hunting activities. The lion usually only stops eating once its stomach is full of meat. Digestion requires further energy expenditure (up to about 25% of total energy expenditure) and thus increases the heat load, forcing the lion to rest. The lion is the only large predator that hunts prey as large as giraffe, buffalo and even elephant.

The leopard also tends to rest for most of the day, in preparation for its usually nocturnal hunting activities. Hunting takes the form of stalking or ambushing prey as the leopard does not spend energy on long chases. The leopard hunts alone and may prey on medium-sized antelope

weighing a little more than its own body weight, which is on average around 60 kg. Leopards will move their kill into the fork of a tree, where the carcass and the leopard are safe from other predators such as hyenas, who would relish chasing the leopard off its kill.

The cheetah, at about 50 kg, is slightly smaller than the leopard and hunts in the cooler morning and early evening, resting in the hottest part of the day. The cheetah is superbly built for pursuit and can reach speeds of close to 100 km/hr. This tempo cannot be kept up for very long, however, and after a few seconds, energy would have to be generated by anaerobic means, leading to a build-up of lactic acid and exhaustion. The cheetah is also limited in its defence abilities, as it has a much weaker jaw than similar-sized predators. It is thus not surprising that the cheetah is only successful in bringing down prey in about 1 in 10 chases and furthermore loses a lot to scavengers, as it is not adapted to defending the carcass. A typical scavenger is the spotted hyena.

Spotted hyenas (about 40 kg) and wild dogs (about 22 kg) are social animals and usually hunt in packs. Both these predators require proportionately more energy daily than the larger cheetah, leopard and lion. The two smaller dog-like predators typically need to feed every day. Impala are probably the main prey and the hyena and wild dog can pursue and isolate an individual as they have sufficient stamina to chase an individual impala for a few kilometres breaking into speeds of over 50 km/hr for short bursts. The members of the pack collectively pursue the prey, and individuals alternate in being the lead chaser. The prey often collapses out of shear exhaustion and both the hyena and wild dog may start feeding before their prey is dead.

In the Kruger National Park these five large carnivores collectively consume about 76 000 prey items per year. On average, this is about one prey item per month for each individual predator. The biomass for these large carnivores is in the region of 0.04 tonnes/km^2 which is much lower than the very numerous impala prey which is about 0.25 tonnes/km^2.

The aardvark is seldom seen as it is nocturnal, resting in burrows during the day, as it needs to protect its thinly covered body from the sun. It weighs on average about 60 kg and, surprisingly for an animal of this size, feeds mainly on termites and ants. Its tongue, similar in length to that of the giraffe, is thus adapted to capture up to 50 000 termites and/or ants in a single night. It digs into termite mounds in pursuit of its favourite prey and uses old mounds for its burrow. The aardvark will also search for ants and termites on the ground.

This dependency on termites and ants as its only energy source has its challenges. During extremely hot periods, the termites and ants may be so reduced in number that the aardvark could literally starve to death. Unfortunately for this insectivorous predator, it is also prey to all five, large, meat-eating predators discussed above.

Endothermic predators: Mammals

The smaller carnivores: Less than 20 kg

The smaller predators, just like the smaller herbivores, have a high surface area to body mass ratio and thus require comparatively more food than the larger predators. They, therefore, concentrate on the more abundant, small, prey animals, such as rodents, birds and insects, which require less energy to hunt. It is often possible to search for and find these animals at a leisurely walking pace. As their prey is generally much smaller in size than their own body mass, the actual capture is of short duration and thus comparatively low in energy expenditure. With a few exceptions, the smaller predator group can be divided into two sub-groups, based on size and dietary preference. Examples of the first group are the caracal, jackal, serval, African wildcat, civet and genet. These all weigh more than 2 kg and primarily feed on small vertebrate prey.

Caracal

The caracal preys on dassies (about 4 kg) for about 50% of its energy requirement, while birds and small antelope make up most of its remaining dietary needs.

There are three mammals in this weight group that have an unusual diet, feeding mainly on termites and ants. These are the pangolin and aardwolf, each weighing about 10 kg, and the smaller bat-eared fox, weighing in at about 4 kg.

Aardwolf

Bat-eared fox

These smaller insectivores require proportionately more energy than the much-larger, termite-loving

Pangolin

aardvark discussed previously. The aardwolf and pangolin may each consume over 200 000 termites per night, while the smaller bat-eared fox may easily consume over 3 000 termites and ants per day, along with many other invertebrate prey animals.

Examples of the second sub-group are meerkats (suricates), mongooses, bats and shrews. These animals, weighing less than 1 kg on average, feed mainly on invertebrates and small vertebrates such as lizards. Meerkats and shrews feed almost exclusively on insects such as beetles, the energy-repackaging agents. The meerkat, found in the arid savanna areas, has a cylindrically shaped body about 35 cm long and, weighing a little less than 1 kg, has a remarkably high surface area to volume ratio. Like similarly shaped mongooses, it needs to warm up its body in the sun before foraging in the morning. Having done so, a meerkat may need to forage between five and eight hours a day to get sufficient energy to support its small body, as it also does not store fat.

The shrew, weighing about 12 grams on average, has more of a spherically shaped body. This shape has a more favourable surface area to volume ratio. Even so, because it is so tiny, it also needs to spend considerable time foraging to satisfy its energy requirement. The shrew consumes twice its weight in insects per day and collectively a group can consume close to 7 000 insects per ha per day. Omnivorous rats and mice, feeding on plant matter and insects, also remove a considerable amount of the invertebrate life.

Bats are mainly nocturnal and there are many species that feed on insects. A single bat colony may be responsible for consuming 100 tonnes of insect prey per year. A single bat in the colony may consume over 1 000 mosquito-sized insects per hour and over 6 000 in a single night. An insect diet provides these small predatory animals with all their energy needs, which has been conveniently packaged by these tiny ectotherms.

CHAPTER 16

Endothermic predators: Birds of prey

Birds of prey, as well as other birds which feed on invertebrates, have similar body shape and size ranges, and thus similar energy demands to those described for herbivorous birds in chapter 11. Included in this chapter are the scavenging raptors (vultures), other diurnal raptors (falcons), the secretary bird and nocturnal raptors (owls). Diurnal birds of prey range from the lappet-faced vulture, weighing just short of 10 kg, to the diminutive pygmy falcon, which is only 100 g in weight. The nocturnal owl ranges in size from the giant eagle owl, weighing about 1.6 kg, to the pearl-spotted owl which weighs about 70 g.

Most birds of prey have strong hooked bills, relatively sturdy legs, superior binocular vision and an acute sense of smell. Except for vultures, their feet are designed for gripping prey. Depending on size, birds of prey eat between 5% and 50% of their body weight per day. A black-shouldered kite, weighing about 0.25 kg, eats about 50 g per day, or 20% of its body mass. The larger Cape vulture, weighing about 9 kg, eats about 540 g per day, but this is only 6% of its body mass. The smaller bird has a much larger surface area to volume ratio, which means that it needs relatively more food to satisfy its energy requirement.

Vultures do not require feet that are adapted to gripping, as they scavenge on prey killed by other animals, as well as animals which have died of natural causes.

Their excellent vision enables vultures to locate carcasses up to 2 km away and, in fact, other part-time scavengers (e.g., the spotted hyena) follow vultures to a carcass. Vultures can travel up to 150 km a day in search of carrion and can be responsible for consuming up to 70% of carrion in an area. Travelling long distances in flight is an expensive exercise and large birds such as vultures have evolved energy-saving flying styles. Soaring, where the main propulsive force is the strength of air currents rather than muscular activity, is a relatively cheap locomotory mode. On finding an unpredictable food source such as a fresh carcass, vultures get to work quickly. Different species may arrive in succession. White-headed vultures are often the first arrivals and use their powerful beaks to tear into the carcass. The white-backed vulture then piles in, opening the insides and creating space for other vultures to get to the soft tissues. Finally, the lappet-faced vulture cleans up by feeding on tendons, ligaments and skin. Vultures can strip a carcass in a few hours, thus contributing to a clean and disease-free environment. They are admirably adapted to keeping harmful bacteria, which may be present in vast numbers in decaying carcasses, under control. The absence of feathers on their heads and necks prevents bacteria from accumulating on these areas and any harmful bacteria consumed would be killed by their stomach acid, which is much stronger than that of any other birds. The important ecological role of vultures is carried out by only a few species. Most savanna areas have only about five resident species and the population numbers of these are relatively low. To a large degree they depend on the relatively small amount of energy left over in animal carcasses, after these have been fed on by the large mammalian predators. The small vulture populations suffer from being poisoned, often outside of protected areas, as they scan vast areas for carcasses and this includes moving across protected-area boundaries.

The diurnal raptors, such as eagles, hawks, kites, falcons and buzzards, employ a variety of hunting techniques and many will feed on other birds. The crowned eagle, a powerful raptor, may even feed on small antelope and monkeys, while the favourite food of the black eagle is the dassie.

The snake eagles, along with the bateleur, commonly prey on snakes. They have relatively small feet but have rough spines underneath the feet which assist in gripping squirming snakes. Falcons are adept at preying on other birds in the air, while hawks often perch on trees and chase birds through the tree canopies. Some kites and kestrels are able to hover, especially when flying into the wind, thus being relatively stable in the air while they scan for prey. Terrestrial birds such as francolin and guineafowl are a common part of the diet of many of the above raptors.

Somewhat of an oddity is the secretary bird, which walks along the ground hunting for prey. It strikes rapidly, confusing its prey, and then stamps on it. It even takes snakes in this manner.

The nocturnal species, the owls, have some special adaptations for hunting in the dark.

Much of the skull area is taken up by

exceptionally large eyes, which are rich in light-sensitive cells. Their ears are also large and the parts of the brain that control these functions are particularly well developed. Although owls have binocular vision, the structure of their eyes restricts movement within the eye sockets, and they therefore have a rather narrow field of vision. They compensate for this by being able to rotate their heads up to 270 degrees. They are superb hunters, many commonly feeding on rodents, but they take whatever suitable-sized prey available. Owls have a large wing area in relation to their body size and soft feathers allow them to fly silently. This allows them to easily pick up any sounds from potential prey and approach them undetected.

The above adaptations enable birds of prey to obtain the high quality of energy from the protein and fat found in meat. They are thus able to obtain enough energy for the requirements of flight, despite their relatively high surface area to body mass ratio.

CHAPTER 17

Endothermic predators: Insectivorous birds

Insectivorous birds are in the main small, weighing less than 100 g and in many cases less than 10 g; many of them are also passerines (birds that perch), which have a metabolic rate some 60% higher than that of most other birds and mammals. This is because they are very small and most probably a consequence of needing to use a lot of energy each time they leave their perch in pursuit of prey. The cost of the initial stages of flight are particularly energy demanding. The high metabolic rate means that they need a relatively constant food supply to satisfy their energy requirement. Luckily, insects are found in vast numbers in the savanna ecosystem and the thousands of different species occupy a large variety of niches exploited by insectivorous birds. In addition, these small birds have many adaptations that help them to overcome the challenge of obtaining sufficient food energy.

The insectivorous birds include aerial feeders, tree bark and ground probers, omnivores, scavengers and many passerines that search for insects in woody and grassland environments.

The aerial feeders

Swifts, martins and swallows are diurnal aerial feeders with long wings, short bills and a wide gape. Their streamlined bodies, slender wings and the ability to glide are adaptations that assist in reducing their metabolic rate. For example, they use 40% less energy in flight compared to other similarly sized passerines. Nevertheless, these aerial feeders still need a considerable amount of energy, so they consume food equivalent to 50%, or more, of their body weight per day. This could mean capturing as many as 900 flying insects per day per bird. When food is in short supply or temperatures drop considerably, the swifts, martins and swallows may enter a state of torpor, which is a period of inactivity accompanied by a drop in body temperature. These small endotherms thus give up the advantage of a regulated body temperature but save an enormous amount of energy. Torpor is particularly advantageous to small endotherms; not only can they cool down relatively quickly, but they can also return to a temperature that allows full activity within a short space of time, often less than 30 minutes. This short duration torpor thus saves on food energy that would be required for higher metabolism if body temperatures remained at regulated higher levels.

The bark, ground and animal body probers

Woodpeckers and hoopoes, larger than most of the aerial feeders at about 65 g, are less streamlined and, instead, are adapted to probing for insects and other invertebrates. Woodpeckers have long, straight bills and feet adapted to clinging to tree trunks, where they may be seen probing the bark for insects. Hoopoes have curved bills which they use to probe for insects living below the soil surface.

Woodpecker   Hoopoe

The oxpeckers, weighing about 50 g, cling to the backs of some herbivores and probe for ticks and other parasites. Their pointed and laterally compressed beaks are ideally shaped for probing. An adult may consume over 100 ticks per day and 10 times as many larvae. Unfortunately for their hosts, they also utilise blood as an energy source and may, themselves, be considered parasitic. However, the relationship is beneficial to the herbivore as well as the birds, which not only remove ticks, but may also alert their host to possible danger by means of an alarm call.

The perching insectivorous birds

The perching insectivorous birds include larks, pipits, wagtails, drongos, shrikes, cuckoos, tits, babblers, warblers, flycatchers, kingfishers and rollers. The body sizes of these birds correlate with their feeding strategies and energy requirements.

The larger kingfishers and rollers, weighing over 40 g, adopt a sit-and-wait hunting strategy. They scan their environment from a perched position, from where they can spot prey up to 30 metres away. Their bills are proportionately larger than those of the smaller birds, allowing them to prey on larger insects such as grasshoppers and beetles. These insects are often too big to be swallowed whole and so need to be crushed beforehand. An advantage of hunting for larger prey is that fewer insects need to be consumed; thus, time between catches is reduced, saving energy.

The much smaller warblers, often less than 12 cm in length and weighing less than 12 g, adopt a hunting strategy termed gleaning. This process involves moving around continuously in search of prey. As they have small, slender bills, they search for small insects on leaves and swallow their prey whole. Birds between these two sizes, such as drongos and shrikes, have beaks less slender than the warblers, but not as large and sturdy as the kingfishers. They adopt an intermediate strategy, hunting for medium-sized insects from a perched position; after capture, they return to their perch and swallow the prey whole.

The larger, perching insectivorous birds are thus typically sit-and-wait predators, while the smaller ones are more wide-ranging foragers. The middle-sized birds have a strategy between these two extremes. The different strategies adopted by these three size categories reduce competition between the birds because they target insect prey of different sizes.

The omnivores

Many omnivorous birds that utilise seed and fruit food resources also prey on insects. These include the so-called terrestrial birds, such as bustards, korhaans, guineafowl and hornbills, and the smaller barbets, orioles, bulbuls, thrushes, chats, robins and starlings. The smaller birds tend to flit among tree and grass foliage as they seek insects and seeds, while the terrestrial birds forage on the ground; the largest of these birds will feed on small reptiles and mammals as well as insects.

CHAPTER 18

Ectothermic predators: Reptiles

General characteristics of reptilian predators

Over 90% of the savanna reptiles are either lizards or snakes. These reptiles all have scales and are mostly very small ectotherms. They are mostly terrestrial while the few terrapins and the one crocodile species are associated with water bodies. Focus in this chapter will be on lizards and snakes, whose sizes and body shapes typically represent the ectothermic predatory lifestyle. The tortoise was discussed under the ectothermic herbivore section while the carnivorous terrapins, who are largely concerned with waterbodies, are not considered. The extremely large ectothermic crocodile will be discussed as it is regularly seen and has some interesting adaptations that make it special. One of the challenges of the body shapes of reptiles relates to a potential conflict between locomotion and breathing. Lungs, hearts and circulation mechanisms have had to be modified, so that sufficient oxygen gets supplied to muscles. Some reptiles have developed special adaptations that assist in maximising oxygen supply, while others still retain some primitive features, and these are discussed for each group. The predatory reptiles probably match the biomass of all the larger mammalian predators in the savanna: a consequence of their more efficient use of energy from their prey items.

The crocodile

Crocodiles are the largest ectotherms found in the savanna, some growing to a length of 6 m and weighing close to 1000 kg. Unlike most endotherms, these reptiles have no hair or feathers to insulate their large bodies and thus, not surprisingly, are associated with relatively large bodies of water where the temperatures do not fluctuate as much as they do in the terrestrial environment.

Strangely, crocodiles are in some respects more closely related to birds than other reptiles. It is thus not surprising that crocodiles share some of the characteristics of birds. The pulmonary and heart system similarities are particularly interesting. The lungs of a crocodile are structured in a manner that allows air flow to be unidirectional, as in birds. Mammals, on the other hand, have a tidal flow system and air enters and leaves by the same openings, which leaves some stale air in the lungs with each breath. The unidirectional method of air flow has the advantages of removing stale air in the lungs, reduces the cost of breathing, the evaporation of water and heat loss. This energy-saving adaptation gives the crocodile an advantage over other reptiles.

Another similarity that the crocodile shares with birds is a four-chambered heart, while other reptiles only have the more primitive three-chambered heart. The four-chambered heart allows them to deliver more oxygenated blood to muscles and assists them in being able to speed up body heating when required. The structure of the heart also helps in oxygen retention when spending hours under water. The crocodile heart has a valve in it that the crocodile can open or close. When open, blood flow is reduced to the lungs and returned to the body. This allows the animal to drop its heart rate to about three beats per minute, thus saving energy. It is also able to use anaerobic sources of oxygen efficiently.

Crocodiles feed mainly on fish, but also use the water body as an ideal base to ambush terrestrial prey that approach the water's edge to drink. They move quite rapidly on land but can only reach speeds of about 12 km/h and then only for short distances. They even gallop, moving their legs from the normal laterally extended position to an almost vertical stance under the body. This is an advantage they have over lizards. Usually, the side-to-side movement of the trunk of tetrapod reptiles restricts lung ventilation (see further on for lizards) but crocodiles have overcome this limitation by changing the volume of the trunk to move air in and out of the lungs. This is achieved by the movement of the liver and kidneys, and by rotation of the pubic bones. The crocodile can also be motionless for up to 2 hours (sit and wait for prey) but a disadvantage is that any exertion rapidly increases lactic acid levels (anaerobic respiration), and the crocodile would need to rest after a short time.

Once it has a chunk of meat of suitable size, the crocodile swallows this whole and has a digestive system that produces highly concentrated hydrochloric acid, which breaks down bones, hooves, etc. The crocodile will also feed on small prey such as birds and lizards.

So, although the crocodile is relatively large for an ectotherm, it is well adapted to using the advantages of a water habitat and can bide its time by being a sit-and-wait predator, and its well-developed ventilation mechanism and cardiac system assist in reducing the expected high-energy requirement of such a large ectotherm. Furthermore, its digestive system is such that it can break down unwanted items such as bones and hooves.

Like most reptiles, the active temperature range for the large crocodile is about 33°C, a few degrees below the steady body temperature maintained by most endotherms (mammals and birds). The crocodile thus does not have to increase body temperature as much as most endotherms do to be active. Their flattened body shape allows them to expose a high proportion of surface area to radiation from the sun and obtain heat through conduction from surfaces, such as rocks and sand. The latter would be taking advantage of the boundary effect, where air just above a surface is considerably hotter, and therefore a good conductor.

The lizards

The 400 species of lizards in southern Africa are much smaller than the crocodile, with over 80% of lizard species weighing less than 100 g and about one-third weighing less than 1 g. Although the energy requirements of these small animals are high because of the relatively high surface area to body mass ratio, they are very numerous. As ectotherms their actual energy requirements would only be one tenth of similar-sized endotherms; as explained in chapter 2, the latter cannot survive at such small sizes.

A second important difference is related to body shape. A body shape that increases surface area to body mass ratio would be a disadvantage to an endotherm. It is thus not surprising that body shapes such as an elongated and/or dorsiventral form are generally absent in endotherms but found in many lizards. The lizards, as ectotherms, cope with this flexibility in body form which would not be possible for small endotherms. This allows the small ectotherms such as lizards to occupy ecological niches that are not available to small endotherms.

One of the disadvantages of the general lizard body shape is that there is a certain amount of conflict between the need for sustained locomotion (e.g., in hunting) and the requirement for a regular supply of oxygen. In mammals the conflict is solved because of the diaphragm muscle, and the fact that the thoracic vertebrae are located above the lungs, which allows a running animal to maintain an adequate oxygen supply. The galloping antelope diagram below indicates that when the vertebral column is bent, pressure on the lungs increases, forcing air out. The second diagram indicates a straightening of the vertebral column, which results in pressure on the lungs falling and air being pulled into the lungs.

Air leaves lungs

Air enters lungs

The situation is different in lizards because the body moves from side to side during locomotion.

The limbs are used in alternate pairs and supply purchase on the substrate as the trunk muscles move the animal. This mode of locomotion is, however, only feasible over short distances. This is because the bending of the trunk also compresses the ribcage to ventilate the lungs and these actions cannot happen together. One lung is compressed as the other expands and air flows from one lung to the other, thus interfering with flow of air into and out of the mouth as the lizard moves.

Sustained locomotion is thus compromised because the lizard would soon have to switch to anaerobic respiration and, as has previously been pointed out, this can only be of short duration. The conflict as outlined below puts some limitation on the hunting abilities of lizards.

Diagram above shows how air is shunted between lungs as lizard moves forward

Most lizards feed on insects and other invertebrates, but will also occasionally prey on other lizards, small birds and mammals. They adopt two main strategies for capturing prey: the sit-and-wait and wide-foraging strategies. Sit-and-wait lizards tend to have a stocky body shape and short tails.

They may remain motionless for hours, waiting for a suitable prey to come within striking distance. By being motionless for long periods (90% of time) these lizards need only a relatively low input of food energy and have a low energy expenditure per day. They rely mainly on anaerobic respiration for the short dashes that are needed to catch prey. Using the glycogen stored in the muscles, ATP (adenosine triphosphate) is manufactured very quickly, but lactic acid is produced as an end-product, which prevents any further metabolism and forces the lizard to rest.

Sit-and-wait lizard

Wide-foraging lizard

At the other extreme, the widely foraging species tend to have a more streamlined, longer body and usually a long tail. This shape allows them to move relatively easily through vegetation as opposed to the bulky sit-and-wait lizard shape, which would hamper movement. These lizards sustain locomotion for up to 80% of the time and therefore have a much higher daily energy expenditure than their sit-and-wait counterparts. They tend to have larger hearts and their blood contains more red blood cells; they are thus able to utilise aerobic respiration for longer periods than the sit-and-wait lizards. So, although the daily energy expense for the widely foraging lizards can be one and a half times greater than the sit-and-wait species, they adequately compensate for this by consuming about twice as much food energy. They will therefore have excess energy for growth and reproduction. Both foraging types have high assimilation rates and high production efficiencies, both over 80%. Essentially, the small lizards can catch prey in small spaces not available to most endotherms and thus efficiently repackage the energy from their prey, which is passed on to the predators of lizards themselves.

The above examples focus on shape and foraging method. What about size? It has been shown that smaller lizards can have different environments for foraging than larger ones, even where the prey items are similar. This is also seen in juveniles and adults of the same species. Some larger lizards can only bask in the sun for a relatively short time and must then return to a shady environment. As they are relatively large, they cool down relatively slowly and so can forage in the shade for some time before they need to bask again. The smaller lizards have the opposite strategy, as they forage in the sun and retire to the shade only when they need to cool down. In the shade their temperature drops rapidly and so they need to return to exposed sites to forage. So, it is not only shape that matters, but size as well, even involving juveniles and adults of the same species.

BOX 13: THE EXCEPTIONALLY LARGE LIZARDS

There are two exceptionally large lizards commonly found in the savanna: the rock monitor and the water monitor. They can grow up to 2 m in length and may weigh up to 8 kg, occasionally more. They are voracious predators and, like many other lizards, feed on anything from insects to small mammals. They have forked tongues and are able to use scent particles to follow prey, as snakes do. Another feature they have in common with large snakes such as the python, is that their hearts are slightly better advanced than the typical three-chambered hearts of other lizards. They have a ventricular septum, which allows for oxygenated blood to flow rapidly to the body, while preventing the lungs being flooded by high-pressure blood. This gives them a relatively high aerobic scope which allows them to be active for longer. Crocodiles and pythons are the main predators of the water monitor, while the martial eagle and the honey badger have preference for the rock monitor.

The snakes

There are over 150 species of snakes in southern Africa. Their long, relatively cylindrical bodies are not conducive to a favourable surface area to body mass ratio; however, these ectotherms are adapted to living with this potentially energy-sapping body form. There are a few adaptations that snakes have which allows them to cope with the unfavourable elongated body shape. Firstly, the reduction in diameter associated with elongation in snakes has resulted in some special modifications. The left lung is absent or vestigial, the gallbladder is found behind the liver and the right kidney is placed in front of the left, with the same arrangement for the two gonads. This creates space within the unfavourable body form for large prey and the atrophy of some organs also contributes to metabolic savings. Secondly, the snake jaw is designed to accommodate large prey, as it is very kinetic. Elongated lizards can only swallow smaller prey because of their relatively small mouths, whereas snakes can engulf prey that is considerably larger than themselves. They can even continue breathing while consuming large prey, by extending the glottis outside of the mouth for air intake. A further advantage to the snake is that it breathes by contracting muscles between the ribs (inspiration); when these muscles relax as contraction moves down the body to aid in locomotion, expiration occurs accompanied by the characteristic snake hiss.

The snake has vertebrae right along the length of its body and these are used in association with the body muscles, not only for locomotion and respiration, but also for passing captured prey along the elongated body to the sites of digestion. The digestive enzymes in a snake are immensely powerful and are also found in the saliva with which the snake first covers its prey before swallowing. Actual digestion may take from a few hours to days. Strong enzymes dissolve bones and eggshells, while hair and claws are usually excreted.

Many snakes only feed intermittently, at intervals of a few days to a year or more. When a snake is active, its digestive system is inactive and, in fact, shrinks. The reverse is true when it is digesting. In fact, snakes can increase their heart size during digestion, resulting in more efficient delivery of oxygen which is in high demand during digestion. Assimilation rates are generally above 70% and thus like those of the endothermic mammal and bird predators, but production efficiencies are higher than those of the endotherms.

It is possible to consider the predator activity in snakes in terms of either the sit-and-wait or widely foraging extremes. Large snakes with relatively thick bodies, such as the puffadder (1 m long, 6 kg mass and a girth of 30 cm), adopt a sit-and-wait strategy, while more streamlined snakes, like the mamba and cobra (1.8 m long, 1 kg mass and a girth of about 15 cm), are more active wide foraging species.

These ectotherms thus play a role in ecosystem services in a special way. Their locomotion style, associated with their slender body shape, allows them to contribute to the population control of some animals that other carnivores do not feed on as much. They are, for example, significant predators of the many species of rodents. These successful small endotherm hindgut herbivores occupy small and slender spaces that many predators cannot enter, but snakes can conveniently catch them, as their long and slender bodies allow them to enter the confined narrow spaces. Many of the rodents can have densities of up to 40/ha and a biomass per hectare that can exceed those of grasshoppers or even impala. These densities would be much higher, probably reaching plague numbers, if it were not for the contribution of snake predation.

Puffadder: Sit and wait snake form

Cobra: Wide foraging snake form

CHAPTER 19

Ectothermic predators: Invertebrates

The predatory invertebrates, like those of their herbivorous cousins, comprise incredible numbers of species and individuals. These include insects, spiders, scorpions, solifugids and centipedes, each of which have more than 100 species in the savanna. Collectively, these dedicated carnivores consume prey items amounting to thousands of tonnes, so their role in the ecosystem is immense. Ironically, they are responsible for keeping under control the population numbers of their own kind, as they prey on other invertebrates, mostly smaller than themselves. They may adopt a sit-and-wait or wide foraging hunting strategy.

There are well over 1 000 species of spiders in the savanna, with many millions of individuals, each individual spider probably eating about 10% of its body weight per day. Spiders probably prey on as many insects as those consumed by birds and bats in the savanna. The biomass of spiders in savanna ecosystems probably averages about 0.18 tonnes/km^2, which is some 10 times higher than the biomass of the apex mammalian predator, the lion. It has been shown in grasslands that wide-foraging spider species can reduce grasshopper numbers significantly, thus preventing total removal of grass. The sit-and-wait species mainly use webs to catch prey and in grassland areas, where they dominate, predation on grasshoppers is considerably lower, and there is thus a greater consumption of grass plants by grasshoppers.

Insects, such as antlions, Matabele ants, assassin bugs, ground beetles and mantids, mainly eat other insects. Many insects have limited vision and therefore it makes sense for them to adopt the sit-and-wait tactic. Some may be cryptic, resembling bark or leaves, or they may mimic an object such as a flower, as an aid to ambushing their prey.

The praying mantis is a well camouflaged sit-and-wait predator while antlion larvae simply wait for prey to fall into their cone-shaped pit-trap on the soil surface. The adult antlion is winged and looks very similar to a dragonfly. The adults only live for about four weeks, their sole purpose being to mate and lay eggs in this short period. The larvae are the ferocious predatory part of the life cycle. They patiently sit at the bottom of a cone-shaped pit and capture any small invertebrate that falls into the pit. They have strong mandibles to subdue prey and the mandibles are also hollow so that they can easily suck the body fluids out of the prey. These cones can often be observed in soft sand areas and there can be more than 50 of these 2.5-cm-diameter-sized pits per square metre. The most frequently caught prey are ants, and no doubt these would include Matabele ants.

Matabele ants are wide-foraging hunters that march in

large numbers of up to 2 500 adults. They live in large subterranean nests up to 0.7 m deep and such a colony may contain over one million ants. They predate virtually exclusively on termites. The Matabele ant is one of the largest ant species with a length of about 2.5 cm and comfortably overpowers the smaller termite. They carry the dead termites back to the nest for actual feeding. One of the interesting facts about the ant is that it is the only known ant species that will also carry lightly injured colleagues back to the nest where they are nursed back to health. This process reduces the potential mortality rate from about 32% to almost zero. It has been estimated that through this caring process the colony size can increase by some 29% and is an energy-saving activity, as less energy needs to be spent on reproduction to increase the colony size.

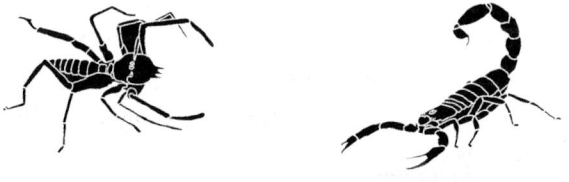

Other wide-ranging foragers are the solifugids and scorpions.

As mentioned before, these invertebrates and the smaller lizards and snakes occupy ecological niches not available to the larger ectotherms or endotherms. They can thus be considered as packaging energy to be used by larger predatory animals higher up the food chain. They also provide an important service to the ecosystem, helping to control populations of insect species, which are not consumed by other predators.

SECTION F

The Decomposers: The Final Consumers

All net ecosystem energy captured in plants that is not returned across the boundaries of the ecosystem through respiration by herbivores, carnivores and omnivores in the grazing food chain will do so by passing through the detrital food chain. About 70% of the primary production is decomposed in this latter process, along with the waste and dead tissues of consumers from other trophic levels.

Insects are by far the most dominant of all animal groups that are responsible for the final breakdown and release of energy from the decomposer system. Even if some of the decomposer organisms are themselves fed on by other organisms (e.g., birds), that captured energy will eventually be released through the decomposition process.

The first chapter in this section covers the general process of decomposition, while the second and third chapters cover decomposition examples from dead plant and animal matter, respectively.

CHAPTER 20

Decomposition: The process

Decomposition is essentially the breakdown of chemical bonds in dead tissue, releasing energy and inorganic nutrients to the environment. Nutrients which have been incorporated (**immobilised**) into plant and animal tissue are **mineralised** and returned to the environment as inorganic material, through the process of decomposition.

Decomposers comprise the greatest proportion of biomass in an ecosystem after the primary producers (plants). Ecosystems could, in fact, function with only the primary producers and decomposers. Probably less than 30% of net primary production in terrestrial systems passes through the herbivore consumer trophic levels, which means that some 70% of net primary production is passed on as dead material directly to the decomposers.

It is not surprising that this massive resource base supports an enormous diversity of species, and their population numbers in the decomposition system are astounding. One cubic metre of savanna soil may contain over 1 000 decomposer species and well over 5 million individual organisms. The species include the so-called microflora (bacteria and fungi), the protozoa (e.g., amoeba) and many orders of invertebrates, including termites, beetles, blowflies, earthworms and millipedes. The microflora are the organisms responsible for the final mineralisation of organic matter and releasing nutrients in inorganic form back to the soil. The other decomposers, the detritivores, are involved in speeding up the mineralisation process by fragmenting tissues into smaller bits, which the microflora can utilise more efficiently.

The decomposition process, or detrital food chain, is often distinguished from the grazing food chain, firstly, because its resource base is detritus (dead organic matter) and, secondly, because of its mineralisation function. However, there is a considerable overlap between the two feeding chains; for example, birds, which are part of the grazing system, feed on earthworms, beetles and termites, which are all part of the decomposer system. Probably the best example that has been covered in this guide is the predation by the aardvark on the decomposer organism, the termite. It is also not surprising to find that within the decomposer system itself, there are organisms that are specialist carnivores, such as spiders and centipedes. During these activities, organic matter that has already entered the decomposition process is immobilised for a second time and will only re-enter the decomposition process on the death of these predators themselves.

Four examples of decomposition are discussed, two from the primary producer level (dead leaves and plant stems) and two from the animal consumer level (dead animals and faeces).

CHAPTER 21

Decomposers of dead plant parts

Leaf decomposers

Decomposition begins when leaves senesce and produce sugary secretions, which are fed on by microflora (bacteria and fungi) commonly found in the air and in water. This process continues and accelerates as the leaf tissues are invaded by a succession of specialist microflora. The early arrivals are the microflora that utilise soluble materials such as sugars, and these early colonists, are often referred to as the 'sugar' bacteria and fungi. They must, however, be replaced by microflora species that are specialists at decomposing more resistant components such as starch and cellulose. The process of decomposition slows down as the more resistant compounds are tackled. It is the diversity of species that allows the complex leaf tissues to be broken down completely.

The process is accelerated as dead leaves fall to the ground and fragmentation of tissues by the detritivores, such as earthworms, millipedes and harvester termites, begins. This creates a greater surface area and broken cells for the microflora to mineralise more efficiently. The detritivores can increase the surface area available to the microflora by up to 15 times. The detritivores themselves incorporate some of the leaf material into their own growth, as do the microflora. This is a further example of nutrient immobilisation, because the nutrients contained within the litter are not immediately available for uptake by plants. Assimilation by detritivores is not very efficient, though, only about 10%, so a lot of leaf material is returned to the soil through the gut as faeces and can then be colonised by the microflora for mineralisation. Microbivores (e.g., mites, beetle larvae) feed on the bacteria and fungi, and detritivores (e.g., harvester termites) are fed on by animals typical of the grazing food chain as described above. Eventually, though, all this captured energy will be released by decomposition of the consumers responsible for immobilising nutrients for a second time.

Plant stem decomposers

Fungi are responsible for the eventual mineralisation of dead plant stems (grass and wood). These plant parts have extremely high concentrations of structural organic compounds such as cellulose and lignin. The decomposition process is, therefore, much slower than for the softer leaves, because the fungi must first penetrate the highly lignified stems. Specialist fungi are found: the brown rots which decompose cellulose and the white rots which mainly decompose lignin. The detritivores are again involved in speeding up the process through fragmentation. Emphasis will be placed on the termites (about 200 species), although there are many other species of detritivores, such as wood borers, associated with wood decomposition.

Termites

Most of the termite species live in underground nests, but only some species build the earth mounds that are so characteristic of the savanna bushveld. These mounds serve as air-conditioning systems which assist in keeping the underground nest areas at a favourable temperature. The bottom of each nest area is well connected to a water source, which helps with controlling humidity in the nest. In the savanna there is an average of about one termite mound per hectare, but in some areas they may be found at even greater densities. One mound may contain over 200 000 individuals, responsible for removing about 60% of annual wood fall and about 5% of leaf litter around the mound; together this could total about 200 kg/ha each year. They are thus extremely important, performing a strategic ecological service in terms of nutrient cycling.

One group of termites has a mutualistic association with protozoa for digestion purposes. The protozoa are found in the hindgut of the termite, where they ingest fine particles of plant stems which have passed through the gut after fragmentation by the termite mandibles. The protozoa are very efficient in decomposing cellulose, but less so in breaking down lignin. Specialised fungi do this either in the gut or after deposition of faeces. These termites, in fact, re-ingest their faeces so that there is a second attack on the cellulose and lignin.

Another group of termites, which includes the mound builders, has evolved to employ an even more advanced technique. They farm fungus gardens, where fungi break down cellulose and lignin outside the termite bodies. Workers of this macro termite group collect and chew stems and, on returning to the nest, defecate the fragmented stem particles, which are then called combs. Soon the combs are invaded by specialised fungi which break down the cellulose and lignin to simpler sugars. These are then eaten by workers and used for their own growth; they are then passed on to other members of the colony, including the queen. Eventually, on the death of a termite, organic matter would be mineralised and inorganic nutrients returned to the soil.

Termites are a source of food for many birds, mammals and reptiles. As mentioned before, the aardvark can consume as many as 50 000 termites in one night. The termite queen can produce up to 30 000 eggs in a day, compensating in part for those that have been consumed by the aardvark and other predators.

An indirect benefit of termite mounds is that, when inactive, they are used for shelter by at least 20 larger animal species, such as hyena and warthog. Many trees, such as the jackal berry, grow on old mounds which are rich in nutrients and provide protection from fire due to their elevation. Trees are also able to obtain moisture through the various channels created by the termites through the soil below the nest. These micro-habitats eventually support a healthy community of shrubs and trees, which create suitable niches for many animals.

CHAPTER 22

Decomposers of the bodies and waste products of consumers

Decomposers of animal tissue

Vertebrate scavengers, such as vultures and spotted hyenas, may remove up to 90% of a large animal carcass. The remainder of the carcass and those carcasses that may have been missed by scavengers, are subject to immediate decomposition by detritivores and microflora. Unlike plant material, animal carcasses have a higher percentage of protein. This means that decomposition can proceed faster than that for plant material, as nitrogen is not in short supply.

Blowflies (some 150 species in southern Africa) lay their eggs on or in the dead animal and the resulting maggots get to work on the carcass. These maggots are voracious eaters and move in masses which could be up to 500 000 maggots per kg of carcass. Their mouths have hooks with which they burrow into the flesh, breathing through spiracles in their rear ends. As they burrow, they secrete digestive enzymes and spread putrefying bacteria, which continue the mineralisation process.

The resulting smell can be unbearable, particularly in hot weather. A medium-sized antelope carcass could be consumed in this manner within a few days. After feeding to their full, the maggots drop to the soil, pupate and eventually turn into adult blowflies.

Microflora, particularly bacteria, are always present in the environment, but thrive in cooler conditions when the maggots struggle to dominate. There are many specialist microflora species which mineralise bone, hair, skin, sinew, etc. Energy is released by microflora during respiration and inorganic nutrients are returned to the soil.

Decomposers of dung

Dung beetles are the most important decomposers of dung. The accumulation of animal faeces would be a major disruption to the savanna ecosystem and it is therefore essential that these wastes are decomposed. The fly problem would become immense if there were large deposits of dung available for breeding purposes. Furthermore, if dung is not removed, soil compaction occurs, and plants covered by dung become, at least temporarily, unacceptable to herbivores. In fact, undecomposed dung creates a bottleneck in the recycling of nutrients, as only about 20% of faecal nitrogen is returned to the soil; this can lower the productivity of plants for up to a year. When dung beetles are present, they bury up to 90% of faecal nitrogen and thus higher nutrient uptake occurs, and increased plant yield is possible.

BOX 14: THE AFRICAN DUNG BEETLE IN AUSTRALIA

One of the best examples of the lack of dung removal comes from the Australian experience of introducing cattle to the country. The first cattle, numbering seven, were brought to Australia about 240 years ago. Some two hundred years later the cattle population had ballooned to about 30 million, collectively producing some 300 million dung pads a day, which covered the equivalent of 4 million hectares per year. The Australian dung beetles were specialised for kangaroo dung and could not decompose the cattle dung. The fly problem became immense and huge areas suffered from a lack of grass productivity. It was thus decided to introduce cattle dung beetles from Africa, in the hope of reducing the problem. The introduced dung beetles were sterilised but this, ironically, destroyed a micro-organism which enabled the dung beetles to navigate. It was only after the introduction of non-sterilised beetles that they proceeded to bury the dung, thus reducing the fly problem and contamination of natural grasslands by dung pads. This is a good example not only of the specialisation of detritivores, but also of symbiotic relationships necessary for proper functioning and the consequences of massive disturbance in a natural ecosystem.

Southern Africa has over 1 000 different species of dung beetle and they contribute greatly to the maintenance of ecological balance. One tonne of dung can be buried per hectare per year in the savanna. Population numbers are exceptionally high; for example, one pile of elephant dung may house over 15 000 beetles. The many species can be divided into four groups, based on the manner of handling dung. The rollers are the group most likely to be spotted on the road, as they roll a ball of dung away from its original place of deposit to be buried or eaten elsewhere. The stealers capture dung balls from the rollers for their own use. A third group lives within the dung pile and are referred to as dwellers. The tunnellers, on the other hand, bury dung under the pile, where it provides food for beetle larvae.

About 75% of dung beetle species utilise only herbivore dung, with different species specialising in dung from different herbivore species. Carnivore dung is of poor quality, mainly because carnivores assimilate their food extremely efficiently (80%+ is digested), so that their faeces contain only the least digestible components. Carnivores are, of course, much less numerous than herbivores and therefore produce less dung for the use of specialist detritivores. Most carnivore dung may, in fact, be decomposed almost entirely by fungi and bacteria.

SECTION G
Reflections

This section brings together all the information from the examples used in the previous sections. It essentially considers the use and flow of energy in the ecosystem context. It emphasises the importance of size and shape of plants and animals and their thermoregulatory mechanisms to understanding bushveld ecology. This single chapter is thus a reflection of the approach I use in this guide.

CHAPTER 23
Consolidating the theme

In the preface, I refer to the concepts of rarity and commonness and will reflect on these, in relation to energy use and its flow through the savanna ecosystem, from the primary producers to the decomposers and beyond.

Most visitors to the savanna focus on the endotherms (mammals and birds), but from an ecological point of view, the ectotherms (plants, reptiles and invertebrates) are far from unimportant. Throughout this guide, I have emphasised the importance of size, shape and thermoregulation mechanisms of plants and animals, and all these criteria influence energy use and transfer in both endotherms and ectotherms.

Only about 30% of net plant production (primary production) is utilised by herbivores in the grazing/browsing food chain, the other 70% being processed by decomposers through the detrital food chain. Tree and grass leaves make up most of the 30% that is consumed by herbivores. The herbivores that consume most of the plant production are mammals, birds and insects. The first two groups are endotherms and insects are small ectotherms. The consumption, use and transfer of energy by these groups are thus a reflection of my approach which focuses on thermoregulatory mechanisms and size and shape of organisms. Table 23.1

summarises net plant production energy units consumed and subsequently used, by these three important herbivore groups, using arbitrary values. The arbitrary energy units indicate the relative importance of endotherms and ectotherms, regarding energy capture and flow at the primary consumer level of the ecosystem. The use of the 30 available energy units (equivalent to the 30% mentioned above) is allocated to insects (small ectotherms, feeding on leaves, fruits, seeds, etc.), birds (small endotherms using high-energy plant parts) and mammalian herbivores (variously sized endotherms using mainly browse and graze). The percentages used for the intake, assimilation and growth calculations for each herbivore group come from information provided in the relevant chapters. They are used for illustrative purposes, rather than as definitive values.

Table 23.1. Flow of energy from the net plant production to and through the main grazing/browsing herbivore groups, using arbitrary energy units

	Estimated plant energy units consumed	Estimated % of consumed energy assimilated	Estimated energy units assimilated	Estimated % of assimilated energy used for growth	Energy units available to secondary consumers
Insects	5	60	3.0	53	1.6
Birds	5	70	3.5	1	0.035
Mammals	20	50	10	13	1.3
Total/ average	30	60	16.5	22	2.935

Insects and birds consume the same portion of net plant production, birds consuming high-energy plant parts (nectar, fruit and seed) while insects feed on these plant parts plus leaves. The many small ectothermic insects have relatively high assimilation rates and particularly high production efficiencies thus contributing to a relatively high biomass amount that is available for the next trophic level. Birds, on the other hand, must consume high-quality food, but because of the high cost of endothermy and the added energy needs of flight, contribute much less energy to the next level. Mammalian herbivores consume four times as much as birds or insects and their food source is predominantly browse and graze. Mammals, however, have the lowest assimilation rates and also very low production efficiencies. Despite having low numbers of species, they have evolved various strategies for obtaining energy from leaves. These strategies are related to the metabolic mass (size and shape), digestive system (foregut or hindgut) and the choice of graze, browse or a combination of the two. Interestingly the very large and very small mammalian herbivores tend to be hindgut fermenters that usually include the higher-quality browse in their diet. The mid-sized herbivores are largely ruminant grazers such as the wildebeest. There is only one herbivore in this size range that is a grazer but not a ruminant and this is the hindgut fermenter, the zebra. This is surprising because it could be assumed that more species of hindgut grazers would be present to utilise some of the 70% of net primary production that is in fact passed on directly to the decomposer system.

One way of looking at this is to view it as a lost opportunity for the herbivores (primary consumers). For whatever reason, the larger herbivores have failed to evolve species or numbers within a species to take advantage of this 'excess' primary production. The inability of larger herbivores to make use of the 'excess' primary production seems to be a bottleneck in the use of primary production. If only the leaf-eating animals are considered, thus ignoring the many birds and invertebrates that feed on fruits, seeds, nectar, etc., there are probably less than 50 species of vertebrate herbivores in the savanna.

In Table 23.1 above the total estimate of units of energy available to the secondary consumers (2.935) is almost 10% of the 30 units consumed, equivalent to the 10% guideline often used for energy transfer between the grazing food chain trophic levels. The contribution of insects to this is immense, indicating that small animals with ectothermic thermoregulatory mechanisms are crucial to the healthy functioning of the savanna ecosystem.

The contribution of the small ectotherms to re-packaging energy continues in the higher feeding (trophic) levels. The predatory insect numbers are exceedingly high and the many small reptile carnivore species (lizards and snakes) are also efficient in repackaging energy for consumption by other predators.

Many of the animal predatory groups have similar tactics for capturing prey. Invertebrates, reptiles and many of the smaller birds adopt either a sit-and-wait or wide-ranging strategy for capturing prey. The wide-ranging strategy is energetically expensive but this challenge is compensated for by capturing more prey. Insects that are sit-and-wait predators and save considerable energy by being sedentary are, however, more likely to be found by wide-ranging lizard species, while these lizards themselves are more likely to be caught by stumbling into a sit-and-wait snake predator. A sit-and-wait snake might then itself be caught by a secretary bird. Thus, the differences in size, shape and feeding mechanism influence not only the amount of energy required but also the strategy adopted to ensure that sufficient energy is obtained. Essentially all the thousands of ectotherm invertebrate species at the secondary consumer level and beyond are energy re-packaging agents. The vertebrate species richness in terrestrial habitats of savanna is in fact incredibly low in comparison, comprising only about 150 mammal species, 550 bird species and possibly 120 reptile species. In this sense the larger endotherm vertebrates that one is more familiar with in

Consolidating the theme

savanna areas are rare in terms of number of different species and individuals. The apex endothermic mammalian predators, although low in numbers and responsible for only a small amount of energy flow, also provide excellent services by influencing the numbers and distribution of the mammalian herbivores in the ecosystem. However, only a relatively small proportion of energy flow in the ecosystem passes through the vertebrate members in the food chain. Most of the energy captured as gross primary production is returned (no gain or loss of energy) across the boundary of the ecosystem through respiration by three groups: plants (primary producers), invertebrate primary and secondary consumers, and the decomposers.

It is thus not surprising to find that there are many more species in these three biological groups. There are over 2 000 plant species contributing to primary production in the savanna and thousands of animal species in the decomposer level. The former are responsible for capturing energy for incorporation into the terrestrial ecosystem while the latter are responsible for ensuring nutrients are made available in inorganic form for future plant uptake.

Regardless of the number, types and populations of the different species occupying the various trophic levels, the energy captured by plants in an ecosystem during photosynthesis is eventually returned to the atmosphere across the ecosystem boundary by the process of respiration. No energy is gained or lost, and yes, size does matter in this process.

It is important to define the boundaries of an ecosystem when considering energy gain and loss. Some parts of ecosystems might for example have definite physical non-natural boundaries (e.g., fences), while other parts may be defined by a river and/or a political boundary (as in the Kruger National Park).

The boundaries of the major ecosystems at a regional level, such as for savanna described in this guide, are defined in relation to the broad climatic pattern in various zones. These are not definite boundaries as it is impossible to precisely determine the cut off lines for the effect of temperature and rainfall. Even when using the two functional types, broad-leaved and fine-leaved savanna, it is not easy to define the boundaries between the two and there are a lot of areas of mixed types.

Many savanna wildlife properties do, however, have definite boundaries and need to be managed carefully because they are often too small for natural ecological processes to manifest themselves. Appendix 1 provides some examples of the management decisions that need to be made in savanna wildlife enterprises that have defined boundaries.

APPENDIX 1

Determination of ecological capacity of defined savanna ecosystems

Management of reserves, whether large or small, needs to consider the ecological carrying capacity of the defined ecosystem. Ecological carrying capacity is based on the sustainable forage of the grass and browse available to support herbivores. This suggests a balance should be maintained between the forage available and its utilisation by herbivores, at least on a yearly basis. The type and number of herbivores that can be supported in a defined ecosystem are key in this endeavour.

It is generally accepted that the smaller the defined ecosystem the greater the need for active management. Even in a relatively large, protected area such as the Kruger National Park it was deemed necessary in the past for culling operations to be implemented. For example, the annual removal of perceived excess numbers of elephants took place until recently. Relatively new management approaches, such as closing of artificial water points, have since been introduced which hopefully will contribute to a natural control of elephant numbers and thus contribute to the natural sustainability of the ecosystem.

The tremendous increase in the number of much smaller reserves and game ranches in the recent past has encouraged an increased interest in ecological carrying capacity. Almost half of the 10 000 game ranches in South Africa occur in savanna areas. The average size of these game ranches is about 1 000 ha, with about 30% of these being larger than the average size. The owners of these ranches would, in the first instance, like to know what type of herbivore game species and how many of each can be stocked on the ranch to maintain the ecological balance.

An understanding of sustained plant production is the key to determining the ecological balance, and the quantity and quality of forage (mainly grass and trees) should be assessed. If this has been achieved, it is then possible to determine the type and numbers of mammalian herbivores (primary consumers) that can be supported within the defined boundaries of the reserve. This is no easy task as primary production can vary from year to year and seasonally; various herbivores may switch between browse and graze, and the social and behavioural characteristics of the herbivores must also be considered. In addition, it must be remembered that a large portion of primary production is utilised by the invertebrate decomposers directly and thus does not pass through the mammalian herbivores. The mammalian herbivores can be referred to as an umbrella guild. If they are adequately in place, it is then assumed that all the smaller organisms can find niches under the umbrella. Furthermore, many of the smaller organisms, such as birds, reptiles and insects, can easily move across the definitive ecosystem boundary, so it makes sense to concentrate on the mammalian herbivore species and numbers in the first instance. Scientists are continuously working on various methods and processes that may contribute to making management decisions as easy as possible. Details of various approaches are not discussed here. The simulated examples

Determination of ecological capacity of defined savanna ecosystems

presented concentrate on the energy requirements of herbivores that could possibly utilise the sustainable forage availability.

The energy requirements of wildlife herbivores have been based on the management of large stock in the agricultural sector. The agricultural approach uses the idea of animal units (AU) per hectare with a large animal unit equated with a steer of 450 kg, which increases its weight by 500 g a day on forage that has a digestibility energy of 55%. This method has been adapted to wildlife herbivores by the introduction of the animal grazing unit (GU) and the animal browsing unit (BU). The standardised GU is the metabolic equivalent of a blue wildebeest weighing 180 kg, while a BU is equivalent to a 140 kg kudu. All herbivores can be assigned a GU and/or BU value based on the proportion of graze and browse in their diet and the ratio of their metabolic mass to that of the wildebeest and kudu.

The first example considers a single grazing herbivore found in a 1 000-ha reserve composed of grassland. The second considers a reserve of browse (trees) and graze (grass) and occupied by different herbivore species.

Example 1

An owner of a 1 000-ha grassland property wanted to stock his property with one herbivore species, species X, that weighs 68 kg. He employed a range scientist to advise him how many animals of species X he could stock on a sustainable basis. The scientist determined the total graze energy of the grassland and then, using the metabolic weight of the animal, he calculated that each animal required 3 ha, therefore 333 animals could be catered for sustainably on the 1 000-ha property. For those who may be interested in the details of the calculations, information is supplied in Box 15.

Box 15: Determination of numbers of species X that can be sustainably placed on a 1 000-ha property

A scientist determined the metabolisable grass energy available in one hectare of a 1 000-ha grassland reserve over a full year. The available grass forage was found to be 360 000 g/ha and the average metabolisable energy from this grass was 2.29 Kj/g. The total metabolisable graze energy is thus 824 400 Kj/ha.

The metabolisable energy required by herbivore species X is equal to its metabolic weight times 293 Kj/day. The chosen herbivore weighed 68 kg and had a metabolic weight of 23.68 kg ($68^{0.75}$) and thus an energy requirement of 6938 Kj/day (293 x 23.68).

It is now possible to use the above information in a formula which calculates the number of days that one animal can be sustained on one hectare. The formula is:

A = (B x C)/D where, A = Number of animal days, B = Available forage in grams, C = Forage metabolizable energy in Kj/g and D = Forage metabolisable energy required per day by animal.

A = (360 000 X 2.29)/6938

A = 118.8 animal days/ha

This is measured over 365 days, so the number of hectares needed per animal is 365/118.8 which is about 3 hectares per animal. Thus, if the total area available to the animal is 1 000 ha then 333 animals could be stocked on a sustainable basis.

Determination of ecological capacity of defined savanna ecosystems

If a smaller herbivore, half the weight (34 kg) of the previous one, is considered and the formula as presented in Box 15 applied, the results would be as follows. Despite the smaller herbivore being half the weight of the larger one, only 560 animals could be supported sustainably. This is considerably less than twice the numbers of the larger animal (666) that can be sustainably supported on the defined ecosystem (1 000-ha grassland). This again illustrates the importance of size, surface area to volume ratio and metabolic requirements.

The example thus supports the theme discussed throughout this guide where the higher metabolic energy requirements of the smaller animal results in a relatively lower total number that can be supported on the grassland.

Example 2

The situation becomes more complex when browse and graze are part of the ecosystem, and a variety of herbivore species are under consideration. The herbivore species could consist of those that are chiefly unselective grazers, selective grazers, mixed feeders and/or those that are mainly browsers. Even digestive system differences such as whether a species is a foregut or hindgut fermenter may be important.

As in the first example the available forage that can support herbivores in a defined ecosystem must be determined. The graze and browse components are the base for this calculation. The approach identifies the available forage in terms of forage availability of grass in GUs and the available forage of browse in BUs. The details of obtaining these are not discussed any further in this guide. The forage availability is then used in conjunction with the size of the ecosystem (reserve) and the GU and/or BU animal equivalents of herbivore species to get a first approximation of the different herbivore species and their numbers that could be supported in the defined area.

An individual bought a 5 000-ha property in the savanna which contains 139 GUs of forage per hectare and 120 BUs of forage per hectare throughout the year. The total forage GUs is equivalent to 695 000 kg/ha (5000 x 139) and the total forage BUs is equivalent to 600 000 kg/ha (5000 x 120).

The reserve supported eight different herbivore species and their numbers and other important information are shown in table A.1.

Table A.1: The numbers, average mass and digestive system of the eight herbivore species on the 5 000-ha reserve and the relative percentage of graze and browse in their diet

Species	No.	Mass(kg)	% graze	% browse	Digestive system
Nonselective grazers					
Zebra	50	260	93	7	Hindgut
Buffalo	15	520	78	22	Foregut
Selective grazer					
Blue Wildebeest	50	180	87	13	Foregut
Mixed feeder					
Eland	20	460	50	50	Foregut
Impala	100	41	45	55	Foregut
Browsers					
Giraffe	10	830	1	99	Foregut
Kudu	15	140	15	85	Foregut
Steenbok	20	10	34	66	Foregut

The information in table A.1 is used along with the GU and BU equivalents and their metabolic mass to determine the present stocking and this is indicated in table A.2.

As for example 1, those interested in the detail of the calculations can follow the process in Box 16.

BOX 16: THE DETERMINATION OF THE TOTAL ANIMAL GU'S AND BU'S THAT THE RESERVE CAN SUPPORT.

One animal GU is the equivalent to a blue wildebeest weighing 180 kg which needs 3.7% of its body weight as daily forage intake. Therefore, one GU = 0.037 x 180 x 365 days = 2 431 kg/yr. The total number of animal GUs that can be supported is determined by dividing the total forage GU by the standard animal GU, that is 695 000/2 431 which indicates 286 animal GUs can be supported on the reserve.

One animal BU is equivalent to a kudu weighing 140 kg which needs 3% of its body weight in browse forage per day. Thus, one BU = 0.03 x 140 x 365 days = 1 533 kg/yr. The total number of BUs that can be supported is 600 000/1 533 which is 391.

The second requirement is to determine how much of the available GU and BU forage is used by the individual species present. The first calculation needed is to determine the GU and BU equivalents for each of the species. The metabolic mass of each species is determined ($M^{0.75}$) and then divided by the standard metabolic masses of the wildebeest to get GU equivalents and by the kudu to get BU equivalents.

Each of the species equivalents is multiplied by the number of animals of each species and the percentage of graze or browse to determine total GUs and BUs for each species on the reserve.

For the zebra: GU equivalent = $260^{0.75}/180^{0.75}$ = 1.3, BU equivalent = $260^{0.75}/140^{0.75}$ = 1.6

The total zebra GU equivalents on the reserve is thus: 50 x 0.93 x 1.3 = 60

The total zebra BU equivalents on the reserve is thus: 50 x 0.07 x 1.6 = 6

The total GUs and BUs for each of the eight species is indicated in table A.2.

Table A.2: Total animal GU and animal BU equivalents on the 5 000-ha reserve for eight species of herbivores

Species	GUs	BUs
Zebra	60	6
Buffalo	26	9
Wildebeest	44	8
Eland	20	24
Impala	14	22
Giraffe	1	38
Kudu	2	13
Steenbok	1	2
Total	168	122

The present stocking GUs (168) and BUs (122) on the reserve is thus below the potential of 286 GUs and 391 BUs (see Box 16).

The owner is conservative and decides that he only wants to stock the reserve at no more than 70% of the potential GU and

BU numbers. Seventy per cent of 286 is 200 GUs and 70 per cent of 391 is 274 BUs.

An additional 32 GUs (200–168) and 152 BUs (274–122) can thus be added to the existing numbers. The owner decides to double the number of giraffe and kudu on the reserve and acquires 20 black rhinoceros, a browser, to add to his existing herbivores. The 820 kg rhino, which is virtually only a browser, would add 72 BUs and 3 GUs and coupled with the doubling of the other two herbivores a total 6 GUs and 123 BUs will be added. The new totals for all herbivores will be 174 GUs (6 + 168) and 245 BUs (123 + 122), which are below the set limit of 200 and 274, respectively. The new herbivore composition and their GU and BU equivalents are shown in table A.3.

The owner is satisfied with his decision and is pleased he has a good mix of grazers, mixed feeders and browsers on the reserve. The acquisition of the rhinos and additional numbers of kudu and giraffe should increase the tourist number to the reserve.

Table A.3: GU and BU equivalents of the nine herbivore species on the 5 000-ha reserve

Herbivore Species	Number	GUs	BUs
Non-selective grazers			
Zebra	50	60	6
Buffalo	15	26	9
Selective grazer			
Wildebeest	50	44	8
Mixed Feeders			
Eland	20	20	24

Herbivore Species	Number	GUs	BUs
Impala	100	14	22
Browsers			
Giraffe	20	2	76
Kudu	30	4	26
Steenbok	20	1	2
Black rhino	20	3	72
Total	325	174	245

He appoints a manager and specifically tasks him with monitoring the vegetation, including the effects of fire and herbivores over the next two years. The owner did notice what seemed to be an unusual amount of termite mounds on the reserve and wants the manager to monitor the termite activity and their effect on the forage along with monitoring the herbivore effect. The manager is also tasked with observing the herbivore social behaviour, age structure, sex ratio and reproductive success. The intention is to possibly consider introducing some carnivores in a few years' time in order to increase the tourism potential, but the owner wants to be confident that he has the right mix and quantity of herbivore species in place and then to consider the introduction of appropriate carnivores to ensure that the reserve is ecologically sustainable. Not a simple exercise.

About the author

Bruce McKenzie, now seventy years old, spent his formative years on a farm in the Free State and attended boarding school in Kimberley. His teenage years were Cape Town based and he studied biology at UCT (BSc., BSc Hons (cum laude), MSc and, a PhD in 1984). After completing his MSc in 1978 he lectured in Plant Ecology at UNITRA (now Walter Sisulu University) in Transkei from 1979 to 1985 and completed his Phd on aspects of Transkei ecology during this stay. He developed a keen interest in helping enthusiastic young people to gain the knowledge and confidence to start their career paths. Bruce moved back to Cape Town and lectured at the University of Western Cape (UWC) from 1986 to 1996. He left academia after being professor and HOD in Department of Botany at UWC to become the first Executive Director of Botanical Society of South Africa from 1997 to 2008 (He had served on a few committees and the council of the society prior to this). During the latter period, while based at Kirstenbosch, he served on many advisory committees (e.g. Minister of Forestry's advisory committee, the Board of the South African Institute of Biodiversity, South African chapter of the World Conservation Union (IUCN)) and interacted with environmental assessment professionals and civil society groups. After a successful period at the Botanical Society where he helped the society reach a sustainable financial position, facilitated the society being a leader in the conservation and education fields, and having a national focus, he decided to move back to academia.. He was granted Honorary Life membership in 2014 for his many years of involvement with the Botanical Society.

Bruce had always prioritised promoting conservation within so called "disadvantaged" universities and has supervised or jointly supervised twenty students who completed postgraduate studies and had then been appointed to roleplaying positions, four reaching executive positions within the conservation sector and three becoming university professors. He saw an opportunity to continue and further support young people in the conservation arena and thus applied for a position in the Conservation Department at the Cape Peninsula University of Technology (CPUT). He remained championing young people at CPUT until retirement in 2016. He was primarily responsible for the Extended Curriculum Programme (ECP) in the Department of Nature Conservation and was obviously successful as he was presented the CPUT Excellence in Teaching Award in 2015. Although research was not his prime goal in academia, he has published over 40 scientific and popular publications and many reports since 1980, mostly co-authored with his supervisor and/or with one or other of his post graduate students or peers. He has travelled to Australia, Chile, Jordan, U.S.A and Britain for professional conferences, undertaken consultancy work as far as the Seychelles, and has led ecological tours to nature areas throughout southern and East Africa. He has recently completed a five- year term as a Board member for CapeNature and has continued with consultancy since retirement, mainly on assessing and reviewing projects and programs in the tertiary education conservation sector. Dr McKenzie describes his ecological interests and contributions as, "a jack of all traits in ecology and a master of none" with a focus on building confidence in, and creating opportunities for the youth. Bruce is married to Felicity and they have three daughters, all living in Cape Town.

Further readings

Du Toit, J., Rogers, K., & Biggs, H. (2003). *The Kruger Experience – Ecology and management of savanna heterogeneity.* Washington: Island Press.

Lovegrove, B. G. (2021). *The Living Deserts of Southern Africa.* Vlaeberg: Penguin Random House.

Mills, G. & Mills, M. (2013). *A Natural History Guide to the Arid Kalahari.* Cape Town: Black Eagle Media (Pty) Ltd.

Scholes, R. J., & Walker, B. H. (1993). *An African savanna: synthesis of the Nylsvlei study.* Cambridge: Cambridge University Press.

Tainton, N. M. (ed) (1999). *Veld Management in South Africa.* Pietermaritzburg: University of Natal Press.

Van As, J., du Preez, J., Brown, L. & Smit, N. (2012). *Life & the Environment.* Cape Town: Struik Nature.

Index

Plants
acacia 38
ana tree 40, 42
bushwillow 41
camel thorn 42
jackal berry 164
knob thorn 42
magic guarri 59
mopane 41, 43, 59, 116
red bush willow 37, 38
shepherd tree 45
smelly shepherd 45, 46, 47, 48
sweet thorn 42, 45, 46, 47, 48
umbrella thorn 38, 39, 42
white-trunked shepherd tree 45

Mammals
aardvark 125, 128, 160, 164
aardwolf 129, 130
African wildcat 129
baboon 75
bat 75, 130, 153
bat-eared fox 117, 129, 130
black rhinoceros 40, 89, 90, 97, 98, 185, 186
black wildebeest 85
blesbok 85
blue wildebeest 40, 75, 79, 80, 81, 82, 83, 84, 85, 91, 173, 179, 182, 183, 184, 186
bontebok 85
buffalo 40, 70, 86, 126, 182, 184, 186
bushbuck 92, 93
caracal 125, 126, 129
cheetah 125, 127
civet 129
dassie 95, 129
dik-dik 92, 93
duiker 40, 92, 93
eland 86, 87, 88, 90, 183, 184, 186
elephant 40, 42, 70, 71, 72, 79, 89, 90, 91, 95, 96, 97, 98, 126, 168, 177, 178
elephant shrew 70
gemsbok 40, 85
genet 129
giraffe 40, 42, 60, 75, 86, 87, 98, 126, 128, 182, 183, 185, 186
hare 95, 96
hartebeest 85
hippopotamus 61, 86, 88, 97, 98
honey badger 149
horse 19

impala 91, 92, 102, 127, 128, 151, 183, 184, 186
jackal 117, 129
kangaroo 167
klipspringer 92
kudu 86, 97, 179, 182, 183, 184, 185, 186
leopard 70, 125, 126
lion 70, 122, 125, 126, 153
mice 96
mongoose 130
monkey 75, 135
nyala 91, 92
pangolin 129, 130
porcupine 93, 94, 97
rat 67
rhinoceros 70, 96
roan 85
rodent 67, 79, 94, 95, 96, 136, 151
sable 85
serval 129
shrew 130
southern reedbuck 91, 92
spotted hyena 125, 127, 128, 134, 164, 165
springbuck 40, 91, 92
steenbok 40, 71, 72, 92, 183, 184, 185, 186
suricate 130
tsessebe 85
warthog 93, 94, 97, 164
waterbuck 85
white rhinoceros 40, 89, 90, 97, 98
wild dog 125, 127, 128,
zebra 19, 75, 79, 80, 81, 82, 83, 84, 85, 90, 97, 173, 182, 184, 186

Birds
babbler 139
barbet 102, 103, 140
bateleur 135
black eagle 135
black shouldered kite 133
buffalo weaver 70
bulbul 102, 140
bustard 103, 140
buzzard 135
canary 102
Cape vulture 133
chat 140
crowned eagle 135
cuckoo 139
dodo 18, 19
dove 100, 101
drongo 139, 140
eagle 135
Egyptian goose 100
falcon 133, 135
finch 102, 135
francolin 102, 103, 135
flycatcher 139
giant eagle owl 133
guineafowl 69, 103, 135, 140
hawk 135
hoopoe 138, 139
hornbill 103, 140
kestrel 135
kingfisher 139, 140
kite 135
korhaan 140
kori bustard 99
lappet-faced vulture 133, 134
lark 139
lourie 102

Index

mannikin 102
martin 138
martial eagle 149
mouse bird 102, 106
nightjar 106
oriole 100, 103, 140
ostrich 100, 103, 106, 107
owl 133, 135, 136
oxpecker 139
parrot 102
pearl-spotted owl 133
pigeon 102
pipit 139
pygmy falcon 133
red-billed quelea 103
robin 140
roller 139
secretary bird 133, 135, 174
shrike 139, 140
snake eagle 135
sparrow 18
starling 102, 140
sunbird 75, 103, 105, 106
swallow 138
swift 106, 138
tit 139
thrush 140
vulture 133, 134, 135, 165
wagtail 139
warbler 139, 140
waxbill 100 ,102
white-backed vulture 134
white-headed vulture 134
woodpecker 138, 139

Reptiles
Cape cobra 69, 150, 151
crocodile 61, 121, 122, 141, 142, 143, 144, 149
giant plated lizard 110
leopard tortoise 70
lizard 67, 77, 109, 116, 121, 122, 130, 141, 143, 144, 145, 146, 147, 148, 149, 155, 174
mamba 150
puffadder 150, 151
python 122, 148, 149
rock monitor 121, 148, 149
snake 30, 109, 121, 122, 135, 141, 149, 150, 151, 155, 174
terrapin 141
tortoise 110, 111, 141
water monitor 121, 148, 149

Invertebrates
ant 128, 129, 130
antlion 70, 154
assassin bug 154
beetle 139, 160, 162
blister beetle 115
blowfly 160, 165
butterfly 114, 115
caterpillar 75, 114
centipede 113, 153, 160
cicadas 117
dragon fly 61, 154
dung beetle 166, 167
earthworm 24, 160, 161
emperor moth 116
flower beetle 115
grasshopper 115, 118, 139, 154
ground beetle 154
hawkmoth 115
honeybee 115

katydids 118
locust 118
mantids 154
matabele ant 154, 155
millipedes 113, 160, 161
mites 113, 162
mopane worm 116
nematode 113
praying mantis 154
rhino beetle 70
scorpions 113, 116, 153, 155

seed bugs 115
snail 76
solifugid 153, 155
solitary bees 115
spider 113, 116, 153, 154, 160
termite 24, 115, 128, 129, 130, 155, 160, 161, 162, 163, 164, 186
tick 139
woodborers 162